Statistics and the Evaluation of Evidence for Forensic Scientists

Statistics and the Evaluation of Evidence for Forensic Scientists

C. G. G. Aitken

*Department of Mathematics and Statistics,
University of Edinburgh, UK*

JOHN WILEY & SONS

Chichester · New York · Brisbane · Toronto · Singapore

Other Wiley Editorial Offices

John Wiley & Sons, Inc., 605 Third Avenue,
New York, NY 10158-0012, USA

Jacaranda Wiley Ltd, 33 Park Road, Milton,
Queensland 4064, Australia

John Wiley & Sons (Canada) Ltd, 22 Worcester Road,
Rexdale, Ontario M9W 1L1, Canada

John Wiley & Sons (SEA) Pte Ltd, 37 Jalan Pemimpin #05-04,
Block B, Union Industrial Building, Singapore 2057

British Library Cataloguing in Publication Data

A catalogue record for this book is available from the British Library

ISBN 0471 955329

Typeset in 10/12pt Photina by Thomson Press (India) Ltd, New Delhi
Printed and bound in Great Britain by Biddles Ltd, Guildford and King's Lynn

To

Liz, David and Catherine

Contents

Preface

In 1977 a paper by Dennis Lindley was published in *Biometrika* with the simple title 'A problem in forensic science'. Using an example based on the refractive indices of glass fragments, Lindley described a method for the evaluation of evidence which combined the two requirements of the forensic scientist, those of comparison and significance, into one statistic with a satisfactorily intuitive interpretation. Not unnaturally the method attracted considerable interest amongst statisticians and forensic scientists interested in seeking good ways of quantifying their evidence. Since then, the methodology and underlying ideas have been developed and extended in theory and application into many areas. These ideas, often with diverse terminology, have been scattered throughout many journals in statistics and forensic science and, with the advent of DNA profiling, in genetics. It is one of the aims of this book to bring these scattered ideas together and, in so doing, to provide a coherent approach to the evaluation of evidence.

The evidence to be evaluated is of a particular kind, known as transfer evidence, or sometimes trace evidence. It is evidence which is transferred between the scene of a crime and a criminal. It takes the form of traces—traces of DNA, traces of blood, of glass, of fibres, of cat hairs and so on. It is amenable to statistical analyses because data are available to assist in the assessment of variability. Assessments of other kinds of evidence, for example, eyewitness evidence, is not discussed.

The approach described in this book is based on the determination of a so-called likelihood ratio. This is a ratio of two probabilities, the probability of the evidence under two competing hypotheses. These hypotheses may be that the defendant is guilty and that he is innocent. Other hypotheses may be more suitable in certain circumstances and various of these are mentioned as appropriate throughout the book.

There are broader connections between statistics and matters forensic which could perhaps be covered by the title 'forensic statistics' and which are not covered here, except briefly. These might include the determination of a probability of guilt, both in the dicta 'innocent until proven guilty' and 'guilty beyond reasonable doubt'. Also, the role of statistical experts as expert witnesses presenting statistical assessments of data or as consultants preparing analyses for counsel is not discussed, nor is the possible involvement of statisticians as independent court assessors. A brief review of books on these other areas in the interface of statistics and the law is given in Chapter 1. There have also been two conferences on forensic statistics (Aitken, 1991a, and Kaye, 1993a) with a third to be held in Edinburgh in 1996. These have included forensic science within their programme but have extended beyond this. Papers have also been presented and discussion sessions held at other conferences (e.g., Aitken, 1993, and Fienberg and Finkelstein, 1995).

The role of uncertainty in forensic science is discussed in Chapter 1. The main theme of the book is that the evaluation of evidence is best achieved through consideration of the likelihood ratio. The justification for this and the derivation of the general result is given in Chapter 2. A correct understanding of variation is required in order to derive expressions for the likelihood ratio and variation is the theme for Chapter 3 where statistical models are given for both discrete and continuous data. A review of other ways of evaluating evidence is given in Chapter 4. However, no other appears, to the author at least, to have the same appeal, both mathematically and forensically as the likelihood ratio and the remainder of the book is concerned with applications of the ratio to various forensic science problems. In Chapter 5, transfer evidence is discussed with particular emphasis on the importance of the direction of transfer, whether from the scene of the crime to the criminal or *vice versa*. Chapters 6 and 7 discuss examples for discrete and continuous data, respectively. The final chapter, Chapter 8, is devoted to a review of DNA profiling, though, given the continuing amount of work on the subject, it is of necessity brief and almost certainly not completely up to date at the time of publication.

In keeping with the theme of the Series, *Statistics in Practice*, the book is intended for forensic scientists as well as statisticians. Forensic scientists may find some of the technical details rather too complicated. A complete understanding of these is, to a large extent, unnecessary if all that is required is an ability to implement the results. Technical details in Chapters 7 and 8 have been placed in Appendices to these chapters so as not to interrupt the flow of the text. Statisticians may, in their turn, find some of the theory, for example in Chapter 1, rather elementary and, if this is the case, then they should feel free to skip over this and move on to the more technical parts of the later chapters.

The role of statistics in forensic science is continuing to increase. This is partly because of the debate continuing over DNA profiling which looks as if it will carry on into the foreseeable future. The increase is also because of increasing research by forensic scientists into areas such as transfer and persistence and because of increasing numbers of data sets. Incorporation of subjective probabilities will also increase, particularly through the role of Bayesian belief networks (Aitken and Gammerman, 1989) and knowledge-based systems (Buckleton and Walsh, 1991; Evett, 1993).

Ian Evett and Dennis Lindley have been at the forefront of research in this area for many years. They have given me invaluable help throughout this time. Both made extremely helpful comments on earlier versions of the book for which I am grateful. I thank Hazel Easey for the assistance she gave with the production of the results in Chapter 8. I am grateful to Ian Evett also for making available the data in Table 7.3. Thanks are due to The University of Edinburgh for granting leave of absence and to my colleagues of the Department of Mathematics and Statistics in particular for shouldering the extra burdens such leave of absence by others entails. I thank also Vic Barnett, the Editor of the Series and the staff of John Wiley and Sons Ltd. for their help throughout the gestation period of this book.

Last, but by no means least, I thank my family for their support and encouragement.

C. G. G. Aitken

Series Preface

Statistics in Practice is an important international series of texts which provide direct coverage of statistical concepts, methods and worked case studies in specific fields of investigation and study.

With sound motivation and many worked practical examples, the books show in down-to-earth terms how to select and use a specific range of statistical techniques in a particular practical field within each title's special topic area.

The books meet the need for statistical support required by professionals and research workers across a wide range of employment fields and research environments. The series covers a variety of subject areas; in medicine and pharmaceutics (e.g. in laboratory testing or clinical trials analysis): in social and administrative fields (e.g. for sample surveys or data analysis); in industry, finance and commerce (e.g. for design or forecasting); in the public services (e.g. in forensic science); in the earth and environmental sciences, and so on.

But the books in the series have an even wider relevance than this. Increasingly, statistical departments in universities and colleges are realising the need to provide at least a proportion of their course-work in highly specific areas of study to equip their graduates for the work environment. For example, it is common for courses to be given on statistics applied to medicine, industry, social and administrative affairs and the books in this series provide support for such courses.

It is our aim to present judiciously chosen and well-written workbooks to meet everyday practical needs. Feedback of views from readers will be most valuable to monitor the success of this aim.

Vic Barnett
Series Editor
1995

1

Uncertainty in Forensic Science

1.1 INTRODUCTION

The purpose of this book is to discuss the statistical and probabilistic evaluation of scientific evidence for forensic scientists. For the most part the evidence to be evaluated will be so-called *transfer* evidence. This may be illustrated as follows.

There is a well-known principle in forensic science known as *Locard's principle* which states that every contact leaves a trace. Thus, for example, suppose a person gains entry to a house by breaking a window and assaults the man of the house, during which assault blood is spilt by both victim and assailant. The criminal may leave traces of his presence at the crime scene in the form of bloodstains from the assault and fibres from his clothing. This evidence is said to be transferred from the criminal to the scene of the crime. The criminal may also take traces of the crime scene away with him. These could include bloodstains from the assault victim, fibres of their clothes and fragments of glass from the broken window. Such evidence is said to be transferred to the criminal from the crime scene. A suspect is soon identified, at a time at which he will not have had the opportunity to change his clothing. The forensic scientists examining the suspect's clothing find similarities amongst all the different types of evidence: blood, fibres and glass fragments. They wish to *evaluate the strength of this evidence*. It is hoped that this book will enable them so to do.

Quantitative issues relating to the distribution of characteristics of interest will be discussed. However, there will also be discussion of qualitative issues such as the choice of a suitable population against which variability in the measurements of the characteristics of interest may be compared. Also, a brief history of statistical aspects of the evaluation of evidence is given in Chapter 4.

1.2 STATISTICS AND THE LAW

The book does not focus on the use of statistics and probabilistic thinking for legal decision making, other than by occasional reference. Also, neither the role of statistical experts as expert witnesses presenting statistical assessments of data, nor as consultants preparing analyses for counsel is discussed. There is a distinction between these two issues (Fienberg, 1989; Tribe, 1971). The main focus of this book is on the assessment of evidence for forensic scientists, in particular for identification purposes. For example, in a case involving a broken window, similarities may be found between the refractive indices of fragments of glass found on the clothing of a suspect and the refractive indices of fragments of glass from the broken window. The assessment of this evidence, in associating the suspect with the scene of the crime, is part of the focus of this book (and is discussed in particular in Section 7.4.2).

For those interested in the issues of statistics and the law beyond those of forensic science, in the sense used in this book, there are several books available and some of these are discussed briefly.

'The evolving role of statistical assessments as evidence in the courts' is the title of a report, edited by Fienberg (1989), by the Panel on Statistical Assessments as Evidence in the Courts formed by the Committee on National Statistics and the Committee on Research on Law Enforcement and the Administration of Justice of the U.S.A., and funded by the National Science Foundation. Through the use of case studies, the report reviews the use of statistics in selected areas of litigation, such as employment discrimination, antitrust litigation and environment law. One case study is concerned with identification in a criminal case. This is the concern of this book and the ideas relevant to this case study, which involves the evidential worth of similarities amongst human head hair samples, will be discussed in greater detail later (Sections 4.4.2 and 4.4.5). The report makes various recommendations, relating to the role of the expert witness, pretrial discovery, the provision of statistical resources, the role of court-appointed experts, the enhancement of the capability of the fact-finder and statistical education for lawyers.

Two books which take the form of textbooks on statistics for lawyers are Vito and Latessa (1989) and Finkelstein and Levin (1990). The former focuses on the presentation of statistical concepts commonly used in criminal justice research. It provides criminological examples to demonstrate the calculation of basic statistics. The latter introduces rather more advanced statistical techniques and again uses case studies to illustrate the techniques.

The particular area of discrimination litigation is covered by a set of papers edited by Kaye and Aickin (1986). This starts by outlining the legal doctrines that underlie discrimination litigation. In particular, there is a fundamental issue relating to discrimination in hiring. The definition of the relevant market from which an employer hires has to be made very clear. For example, consider the case of a man who applies, but is rejected for, a secretarial position. Is the relevant population the general population, the representation of men amongst secretaries in the local labour force or the percentage of male applicants? The choice of a suitable reference population is also one with which the forensic scientist has to be concerned. This is discussed at several points in this book.

Another textbook, which comes in two volumes, is Gastwirth (1988a, b). The book is concerned with civil cases and 'is designed to introduce statistical concepts and their proper use to lawyers and interested policymakers. . .' (volume 1, p. xvii). Two areas are stressed which are usually given less emphasis in most statistical textbooks. The first area is concerned with measures of relative or comparative inequality. These are important because many legal cases are concerned with issues of fairness or equal treatment. The second area is concerned with the combination of results of several related statistical studies. This is important because existing administrative records or currently available studies often have to be used to make legal decisions and public policy; it is not possible to undertake further research.

A collection of papers on Statistics and Public Policy has been edited by Fairley and Mosteller (1977). One issue in the book which relates to a particularly infamous case, the *Collins* case, is discussed in detail later (Section 4.3). Other articles concern policy issues and decision making.

The remit of this book is one which is not covered by these others in great detail. The use of statistics in forensic science in general is discussed in a collection of essays edited by Aitken and Stoney (1991). The remit of this book is to describe statistical procedures for the evaluation of evidence for forensic scientists. This will be done primarily through a modern Bayesian approach. It has its origins in the work of I. J. Good and A. M. Turing as code-breakers at Bletchley Park during World War Two. A brief review of the history is given in Good (1991). An essay on the topic of probability and the weighing of evidence is Good (1950). This also refers to entropy (Shannon, 1948), the expected amount of information from an experiment, and Good remarks that the expected weight of evidence in favour of a hypothesis H and against its complement $\bar{H}$ is equal to the difference of the entropies assuming H and $\bar{H}$, respectively. A brief discussion of a frequentist approach and the problems associated with it is given in Section 4.5.

It is of interest to note that a high proportion of situations involving the formal objective presentation of statistical evidence uses the frequentist approach with tests of significance (Fienberg and Schervish, 1986). However, Fienberg and Schervish go on to say that the majority of examples cited for the use of the Bayesian approach are in the area of identification evidence. It is this area which is the main focus of this book and it is Bayesian analyses which will form the basis for the evaluation of evidence as discussed here. Recent work in applying such analyses to legal matters include Cullison (1969), Fairley (1973), Finkelstein and Fairley (1970, 1971), Lempert (1977), Lindley (1977a, b) and Fienberg and Kadane (1983).

Another approach which will not be discussed here is that of Shafer (1976, 1982). This concerns so-called *belief functions*, see Section 4.1. The theory of belief functions is a very sophisticated theory for assessing uncertainty which endeavours to answer criticisms of both the frequentist and Bayesian approaches to inference. Belief functions are non-additive in the sense that belief in an event A {denoted Bel(A)} and belief in the opposite of A {denoted Bel($\bar{A}$)} do not sum to 1. See also Shafer (1978) for a historical discussion of non-additivity. Further discussion is beyond the scope of this book. Practical applications are few. One such, however, is to the evaluation of evidence concerning the refractive index of glass (Shafer, 1982).

It is very tempting when assessing evidence to try to determine a value for the probability of guilt of a suspect, or a value for the odds in favour of guilt and perhaps even reach a decision regarding the suspect's guilt. However, this is the role of the jury and/or judge. It is not the role of the forensic scientist or statistical expert witness to give an opinion on this (Evett, 1983). It is permissible for the scientist to say that the evidence is 1000 times more likely, say, if the suspect is guilty than if he is innocent. It is not permissible to interpret this to say that, because of the evidence, it is 1000 times more likely that the suspect is guilty than innocent. Some of the difficulties associated with assessments of probabilities are discussed by Tversky and Kahneman (1974).

1.3   UNCERTAINTY IN SCIENTIFIC EVIDENCE

Scientific evidence requires considerable care in its interpretation. There are problems concerned with the random variation naturally associated with scientific observations. There are problems concerned with the definition of a suitable reference population against which concepts of rarity or commonality may be assessed. There are problems concerned with the choice of a measure of the value of the evidence.

The effect of the random variation can be assessed with the appropriate use of probabilistic and statistical ideas. There is variability associated with scientific observations. Variability is a phenomenon which occurs in many places. People are of different sexes, determination of which is made at conception. People are of different height, weight and intellectual ability, for example. The variation in height and weight is dependent on a person's sex. In general, females tend to be lighter and shorter than males. However, variation is such that there can be tall, heavy females and short, light males. At birth, it is uncertain how tall or how heavy the baby will be as an adult. However, at birth, it is known whether the baby is a boy or a girl. This knowledge affects the uncertainty associated with the predictions of adult height and weight.

People are of different blood groups. A person's blood group does not depend on the age or sex of the person but does depend on the person's ethnicity. The refractive index of glass varies within and between windows. Observation of glass as to whether it is window or bottle glass will affect the uncertainty associated with the prediction of its refractive index and that of other pieces of glass which may be thought to come from the same origin.

It may be thought that, because there is variation in scientific observations, it is not possible to make quantitative judgements regarding any comparisons between two sets of observations. The two sets are either different or they are not and there is no more to be said. However, this is not so. There are many phenomena which vary but they vary in certain specific ways. It is possible to represent these specific ways mathematically. For example, two such ways are known as the binomial distribution and the Normal distribution, introduced in Sections 3.3.3 and 3.4.1, respectively. It is then possible to assess differences quantitatively and to provide a measure of uncertainty associated with such assessments.

It is useful to recognise the distinction between statistics and probability. Probability is a deductive process which argues from the general to the particular. Consider a fair coin, that is, one in which when tossed the probability of a head landing uppermost equals the probability of a tail landing uppermost, equals 1/2. A fair coin is tossed ten times. Probability theory enables a determination to be made of the probability that there are three heads and seven tails, say. The general concept of a fair coin is used to determine something about the outcome of the particular case in which it was tossed ten times.

On the other hand, statistics is an inductive process which argues from the particular to the general. Consider a coin which is tossed ten times and there are three heads and seven tails. Statistics enables the question as to whether the coin is fair or not to be addressed. The particular outcome of three heads and seven tails in ten tosses is used to determine something about the general case of whether the coin was fair or not.

Fundamental to both statistics and probability is uncertainty. Given a fair coin, the numbers of heads and tails in ten tosses is uncertain. The probability associated with each outcome may be determined but the actual outcome itself cannot be predicted with certainty. Given the outcome of a particular sequence of ten tosses, information is then available about the fairness or otherwise of the coin. For example, if the outcome were ten heads and no tails, one may believe that the coin is double-headed but it is not certain that this is the case. There is still a non-zero probability (1/1024) that ten tosses of a fair coin will result in ten heads. Indeed this has occurred in the author's own experience. A class of some 130 students were asked each to toss a coin ten times. One student tossed ten consecutive heads from what it is safe to assume was a fair coin. More discussion on the interpretation of such a result is given in Section 2.4.2.

The way in which statistics and probability may be used to evaluate evidence is the theme of this book. Care is required. Statisticians are familiar with variation, as are forensic scientists who observe it in the course of their work. Lawyers, however, prefer certainties. A defendant is found *guilty* or *not guilty* (or, also, in Scotland, *not proven*). The scientist's role is to testify to the worth of the evidence, the role of the statistician and this book is to provide the scientist with a quantitative measure of this worth. It is up to other people (the judge and/or the jury) to use this information as an aid to their deliberations. It is for neither the statistician nor the scientist to pass judgement. The scientist's only role in court is restricted to giving evidence in terms of what has been called *cognation* (Kind, 1994); i.e., the problem of whether or not the evidence from two places (e.g., the scene of the crime and the suspect) has the same origin.

The use of these ideas in forensic science is best introduced through the discussion of several examples. These examples will provide a constant theme throughout the book. Consideration in detail of populations from which the criminal may be thought to have come, to which reference is made below, are discussed in Section 5.4 where they are called *relevant populations*. The value of evidence is measured by a statistic known as the likelihood ratio, and its logarithm. These are introduced in Section 2.5.

1.3.1   Blood stains

Example 1.1   A crime is committed. A blood stain is found at the scene of the crime. All innocent explanations for the presence of the stain are eliminated. A suspect is found. His blood group is identified and found to match that of the crime stain. What is the evidential value of this match? This is a very common situation yet the answer to the question provides plenty of opportunity for discussion of the theme of this book.

Certain other questions need to be addressed before this particular one can be answered. Where was the crime committed, for example? Does it matter? Does the value of the evidence of the blood stain change depending on where the crime was committed?

Apart from his blood group, what else is known about the criminal? In particular, is there any information, such as ethnicity, which may be related to his blood group? What is the population from which the criminal may be thought to have come?

Questions such as these and their effect on the interpretation and evaluation of evidence will be discussed in greater detail. First, consider only the blood group evidence in isolation. Assume the crime was committed in the south east of England and that there is eyewitness evidence that the criminal was a Caucasian. Information is available to the investigating officer about the phenotype distribution for the *ABO* blood group in Caucasians there (e.g., from Table 1.1, Stedman, 1985). The information about the location of the crime and the ethnicity of the criminal is relevant. Blood group frequencies vary across locations and among ethnic groups. A suspect is identified. The blood group of the crime stain and that of the suspect match. The investigating officer knows a little about probability and works out that the probability of two people chosen at random having matching blood groups, given the figures above, is

$$0.472^2 + 0.405^2 + 0.092^2 + 0.031^2 = 0.396, \qquad (1.1)$$

Table 1.1   Phenotypic frequencies in S.E. England, from Stedman (1985)

Phenotype	*O*	*A*	*B*	*AB*
Frequency (%)	47.2	40.5	9.2	3.1

(see Section 4.4). He is not too sure what this result means. Is it high and is a high value incriminating for the suspect? Is it low and is a low value incriminating? In fact, a low value is more incriminating than a high value.

He thinks a little more and remembers that, not only do the blood stains match, but that they are both of blood group *B*. The frequencies of blood groups *O*, *A* and *AB* are not relevant. He works out the probability that two people chosen at random are both of blood group *B* as

$$0.092^2 = 0.0085,$$

(see Section 4.4). He is still not too sure what this means but feels that it is a bit more representative of the information available to him than the previous probability, since it takes account of the actual blood groups of the crime stain and the suspect.

The blood group of the crime stain is *B*. The blood group of the suspect is also *B* (if it were not he would not be a suspect). What is the value of this evidence? The discussion above suggests various possible answers.

1. The probability that two people chosen at random have the same blood group. This is 0.396.
2. The probability that two people chosen at random have the same, pre-specified, blood group. For group *B*, this is 0.0085.
3. The probability that one person, chosen at random, has the same blood group as the crime stain. If the crime stain is of group *B*, this probability is 0.092, from Table 1.1.

The relative merits of these answers will be discussed in Section 4.4 for (1) and (2) and Section 6.2 for (3).

1.3.2   Glass fragments

The previous section discussed an example of the interpretation of the evidence of blood grouping. Consider now an example concerning glass fragments and the measurement of the refractive index of these.

Example 1.2   As before, consider the investigation of a crime. A window has been broken during the commission of the crime. A suspect is found with fragments of glass on his clothing, similar in refractive index to the broken window. Several fragments are taken for investigation and their refractive index measurements taken.

Note that there is a difference here from Example 1.1 where it was assumed that the crime stain had come from the criminal. In Example 1.2, glass is transferred from the crime scene to the criminal. Glass on the suspect need not have come from the scene of the crime; it may have come from elsewhere and by perfectly innocent means. This is an asymmetry associated with this kind of evidence. The evidence is known as *transfer evidence*, as discussed in Section 1.1, because evidence (e.g., blood or glass fragments) has been transferred from the criminal to the scene or *vice versa*. Transfer from the criminal to the scene has to be considered differently from evidence transferred from the scene to the criminal. A full discussion of this is given in Chapter 5.

Return to Example 1.2. Comparison has to be made between the two sets of fragments on the basis of their refractive index measurements. The evidential value of the outcome of this comparison has to be assessed. Notice that it is assumed that none of the fragments has any distinctive features and comparison is based only on the refractive index measurements.

Methods for evaluating such evidence have been discussed in many papers, principally in the late 1970's and early 1980's (Evett, 1977, 1978; Evett and Lambert, 1982, 1984, 1985; Grove, 1981, 1984; Lindley, 1977a; Seheult, 1978; Shafer, 1982). These methods will be described as appropriate in later chapters. A knowledge-based computer system, known as CAGE (Computer Assistance for Glass Evidence) has also been developed. It is reviewed in Buckleton and Walsh (1991) as part of a general review of the use of knowledge-based systems in forensic science.

Evett (1977) gave an example of the sort of problem which may be considered and developed a procedure for evaluating the evidence which mimicked the interpretative thinking of the forensic scientist. The case is an imaginary one. Five fragments from a suspect are to be compared with ten fragments from a window broken at the scene of a crime. The values of the refractive index measurements are given in Table 1.2. The procedure developed by Evett is a two-stage one and is described here briefly. It is a rather arbitrary and hybrid procedure. More details are given in Chapter 4. While it follows the thinking of the forensic scientist, there are interpretative problems, which are described here, in attempting to provide due weight to the evidence. An alternative approach which overcomes these problems is described in Chapter 7.

The first stage is known as the *comparison* stage. The two sets of measurements are compared. The comparison takes the form of the calculation of a statistic, D say. This statistic provides a measure of the difference, known as a *standardised* difference, between the two sets of measurements which takes account of the natural variation there is in the refractive index measurements of glass fragments from within the same window. If the absolute value of D is less than (or equal to)

Table 1.2   Refractive index measurements

Measurements from					
the window	1.518 44	1.518 48	1.518 44	1.518 50	1.518 40
	1.518 48	1.518 46	1.518 46	1.518 44	1.518 48
Measurements from					
the suspect	1.518 48	1.518 50	1.518 48	1.518 44	1.518 46

some pre-specified value, known as a *threshold* value, then the two sets of fragments are deemed to be similar and the second stage is implemented. If the absolute value of D is greater than the threshold value, then the two sets of fragments are deemed to be dissimilar. The two sets of fragments are then deemed to have come from different sources and the second stage is not implemented. (Note the use here of the word *statistic* which in this context can be thought of simply as a function of the observations.)

The second stage is known as the *significance* stage. This stage attempts to determine the significance of the finding, from the first stage, that the two sets of fragments were similar. The significance is determined by calculating the probability of the result that the two sets of fragments were found to be similar, under the assumption that the two sets had come from different sources. If this probability is very low then this assumption is deemed to be false. The fragments are then assumed to come from the same source, an assumption which places the suspect at the crime scene.

The procedure can be criticised on two points. First, in the comparison stage the threshold provides a qualitative step which may provide very different outcomes for two different pairs of observations. One pair of sets of fragments may provide a value of D which is just below the threshold whereas the other pair may provide a value of D just above the threshold. The first pair will proceed to the significance stage, the second will not. Yet, the two pairs may have measurements which are close together. The difference in the consequences is greater than the difference in the measurements merits. A better approach, which is described in Chapter 7, provides a measure of the value of the evidence which decreases as the distance between the two sets of measurements increases, subject, as explained later, to the rarity or otherwise of the measurements.

The second criticism is that the result is difficult to interpret. Because of the effect of the comparison stage, the result is not just simply the probability of the evidence, assuming the two sets of fragments came from different sources. A reasonable interpretation, as will be explained in Section 2.5.1, of the value of the evidence is the effect that it has on the odds in favour of guilt of the suspect. In the two-stage approach this effect is difficult to measure. The first stage discards certain sets of measurements which may have come from the same source. The second stage calculates a probability, not of the evidence but of that part of the evidence for which D was not greater than the threshold value, assuming the two sets came from different sources. It is necessary to compare this probability with the probability of the same result, assuming the two sets came from the same source. There is also an implication in the determination of the probability in the significance stage that a small probability for the evidence, assuming the two sets came from different sources, means that there is a large probability that the two sets came from the same source. This implication is unfounded; see Section 2.3.1.

A review of the two-stage approach and the development of a Bayesian approach is provided by Evett (1986).

As with blood grouping, there are problems associated with the definition of a suitable population from which probability distributions for refractive measurements may be obtained; see, for example, Walsh and Buckleton (1986).

1.3.3   DNA profiling

Example 1.3   The human body is composed of many billions of cells. Developments in molecular genetics have made it possible to study the differences between people in parts of DNA that are not involved in coding for proteins (Jeffreys *et al.*, 1985a, b). The relevant parts of DNA occur in all cells that have a nucleus. These include white blood cells, sperm, cells surrounding hair roots and cells in saliva and these are the ones of greatest interest to the forensic scientist.

DNA is composed of two weakly connected strands of molecules that spiral to form a double helix. Each strand of DNA consists of an almost endless chain of four different chemical building blocks or *bases*, often referred to by their initials, A for adenine, C for cytosine, G for guanine and T for thymine. The C on one strand always pairs with the G on its complementary strand and the A with the T. There is a very large number of possible orderings of these so-called base pairs in a lengthy stretch of DNA. A sample of DNA may be cut by a restriction enzyme. The fragments are separated from one another by an electric field, a process known as electrophoresis. Fragments of particular interest are identified by a labelled probe which fuses to the fragments since it has a DNA sequence complementary to them. For a so-called *single-locus probe*, no more than two bands are produced, each of which represents a separate fragment of DNA. The lengths of the fragments are measured by seeing how far they have moved through an agarose gel. The shorter, lighter, bands move further than the larger, heavier bands. The final product is a maximum of two bands represented in an autoradiograph. The distance the band has migrated is measured in base pairs (*bp*) or kilobase pairs (*kb*). The maximum of two bands for a single locus probe correspond to the two alleles, one from each parent, at the particular locus specific to the probe. A single band may be present for a variety of reasons. Further details are given in Chapter 8. For a general discussion see Jeffreys (1993), Gettinby *et al.* (1993), Weir and Hill (1993), Rothwell (1993), Grubb (1993), Findlay (1993), Dawson (1993) and Kaye (1993b).

The two-stage approach described briefly in the previous section is also used in the interpretation of a DNA profile. The DNA profile produced by a single locus probe is a maximum of two bands on the autoradiograph. The two bands correspond to the maternal and paternal chromosomes. The positions of the bands on the autoradiograph, measured in units of basepairs (*bp*) or kilobase pairs (*kb*), provide the evidence whose value is to be assessed.

The two-stage process is analogous to that developed for the glass fragments. The comparison stage compares the band positions of the two sets of measurements. If the difference is less than a pre-specified value (threshold value) a *match* is declared and the second stage is implemented. Otherwise, the two sets of measurements are declared to come from different sources (people). The second stage determines the significance of a declared match by estimating the probability of the two profiles having a difference less than the threshold value, assuming that they have come from two different sources. This is sometimes called the probability of a coincidental match.

There are many papers discussing the evidential value of the outcomes of a DNA profile. A Bayesian approach is discussed in Evett *et al.* (1989b), Berry (1991a) and Berry *et al.* (1992) and a summary of the more elementary aspects of the approach is provided in Chapter 8.

The band positions are measured with error. The positions of the two bands are compared. A match between two bands (one from the crime scene, one from the suspect) is usually declared if the band positions lie within a certain number, say three, standard deviations of each other, where the standard deviation is a measure of the measurement error. If a match is declared, the position of one band (the crime sample) is taken as the relevant band position for the second stage. Reference is made to a relevant population database, the definition of which will be discussed later, as with the blood and glass examples of Examples 1.1 and 1.2 (Section 5.4). The binning stage, as the significance stage of glass analysis is called in DNA profiling, then estimates the relative frequency of the chosen band position in the relevant database. Relative frequencies are stored in intervals (known as *bins*) of bases or kilobases such as may be used to construct histograms. The appropriate relative frequency then provides a measure, in some sense, of the evidential value of the DNA profile. A small relative frequency corresponds to a small probability of a coincidental match. In this situation, as with the glass fragments, there is a strong inference that the source and receptor samples of DNA came from the same source. A large relative frequency corresponds to a large probability of a coincidental match. In this situation there is a weak inference that the two DNA samples came from the same source.

The so-called *match-binning* approach is subject to criticism. First, there is a threshold value, as with the glass fragments. Second, the interpretation of the relative frequency obtained from the binning stage is difficult. Third, the choice of the relevant database and population is also difficult. These criticisms have been addressed in a USA National Research Council report (NRC, 1992) but this report itself has been the subject of criticism (Devlin *et al.*, 1993).

These examples have been introduced to provide a framework within which the evaluation of evidence may be considered. In order to evaluate evidence, something about which there is much uncertainty, it is necessary to establish a suitable terminology and to have some method for assessing uncertainty. First, some terminology will be introduced, then the method for assessing uncertainty. This method is probability. The role of uncertainty, as represented by probability, in the assessment of the value of scientific evidence will form the basis of the rest of this chapter. A recent commentary on so-called *knowledge management*, of which this is one part, has been given by Evett (1993b).

1.4   TERMINOLOGY

It is necessary to have a clear definition of certain terms. The crime scene and suspect materials have fundamentally different roles. Determination of the probability of a match between two randomly chosen sets of materials is not the important issue. One set of materials, crime scene or suspect, can be assumed to have a known source. It is then required to assess the probability of the corresponding item, suspect or crime scene, matching in some sense, the known set of materials, under two competing hypotheses. Examples 1.1 and 1.2 serve to illustrate this.

Example 1.1 continued   A crime is committed. A blood stain is found at the scene of the crime. All innocent explanations for the presence of the stain are eliminated. A suspect is found. His blood group is identified and found to match that of the crime stain.

  The crime scene material is the blood group of the crime stain. The suspect material is the blood group of the suspect.

Example 1.2 continued   As before, consider the investigation of a crime. A window has been broken during the commission of the crime. Several fragments are taken for investigation and measurements made of their refractive indices. These fragments, as their origin is known, are sometimes known as control fragments and the corresponding measurements are known as control measurements. A suspect is found. Fragments of glass are found on his person and measurements of the refractive indices of these fragments are made. These fragments (measurements) are sometimes known as recovered fragments (measurements). Their origin is not known. They may have come from the window broken at the crime scene but need not necessarily have done so.

  The crime scene material is the fragments of glass and the measurements of refractive index of these at the scene of the crime. The suspect material is the fragments of glass found on the suspect and their refractive index measurements.

  Evidence such as this where the source is known and is of a bulk form will be known as *source* or *bulk form* evidence. These fragments of glass will be known as source or bulk fragments and the corresponding measurements will be known as source or bulk measurements, as their source is known and they have been taken from a bulk form of glass, namely a window (Stoney, 1991). In general, only the term *source* will be used when referring to this type of evidence.

  A suspect is found. Fragments of glass are found on his person and measurements of the refractive indices of these fragments are made. Evidence such as this where the evidence has been received and is in particulate form will be known as *receptor* or *transferred particle* evidence. These fragments (measurements) of glass in this example will be known as *receptor* or *transferred particle* fragments (measurements). Their origin is not known. They have been 'received' from somewhere by the suspect. They are particles which have been transferred to the suspect from somewhere. They may have come from the window broken at the crime scene but need not necessarily have done so.

There will also be occasion to refer to the location at which, or the person on which, the evidence was found. Evidence found at the scene of the crime will be referred to as *crime* evidence. Evidence found on the suspect's clothing or in the suspect's natural environment, such as his home, will be referred to as *suspect* evidence. Note that this does not mean that the evidence itself is of a suspect nature!

Locard's principle (see Section 1.1) is that every contact leaves a trace. In the examples above the contact is that of the criminal with the crime scene. In Example 1.1, the trace is the blood stain at the crime scene. In Example 1.2, the trace is the fragments of glass that would be removed from the crime scene by the criminal (and, later, hopefully, be found on his clothing).

The evidence in both examples is *transfer evidence* (see Section 1.1) or sometimes *trace evidence*. Material has been transferred between the criminal and the scene of the crime. In Example 1.1 blood has been transferred *from* the criminal *to* the scene of the crime. In Example 1.2 fragments of glass *may* have been transferred *from* the scene of the crime *to* the criminal. The direction of transfer in these two examples is different. Also, in the first example the blood at the crime scene has been identified as coming from the criminal. Transfer is known to have taken place. In the second example it is not known that glass has been transferred from the scene of the crime to the criminal. The suspect has glass fragments on his clothing but these need not necessarily have come from the scene of the crime. Indeed if the suspect is innocent and has no connection with the crime, the fragments will not have come from the crime scene.

Other terms have been suggested and these suggestions provide a potential source of confusion. For example, the term *control* has been used to indicate the material whose origin is known. This can be either the bulk source form of the material or the transferred particle form, however. Similarly, the term *recovered* has been used to indicate the material whose origin is unknown. Again, this can be either the bulk source form or the transferred particle form, depending on which has been designated the control form. Alternatively, *known* has been used for 'control' and *questioned* has been used for 'recovered'. See, for example, Brown and Cropp (1987). Also Kind *et al.* (1979) used *crime* for material known to be associated with a crime and *questioned* for material thought to be associated with a crime. All these terms are ambiguous. Consider the recovery of a jumper of unknown origin from a crime scene (Stoney, 1991). A suspect is identified and fibres similar in composition to those on the jumper are found at his place of residence. The two parts of the transfer evidence are the jumper found at the crime scene and the fibres found at the suspect's place of residence. The bulk source form of the material is the jumper. The transferred particle form of the material is the fibres. However, the jumper may not be the control evidence. It is of unknown origin. The fibres, the transferred particle form, could be considered the control as their source is known, in the sense that they have been found at the suspect's place of residence. They are associated with the suspect in a way the jumper, by itself, is not. Similarly, the fibres have a known origin, the jumper has a questioned origin.

Material will be referred to as source form where appropriate and to receptor or transferred particle form where appropriate. This terminology conveys no information as to which of the two forms is of known origin. There are two possibilities for the origin of the material which is taken to be known: the scene of the crime and the suspect. One or other is taken to be known, the other to be unknown. The two sets of material are compared by determining two probabilities, both of which depend on what is assumed known and provide probabilities for what is assumed unknown. The two possibilities for the origin of the material which is taken to be known are called *scene-anchored* and *suspect-anchored*, where the word 'anchored' refers to that which is assumed known (Stoney, 1991). The distinction between scene-anchoring and suspect-anchoring is important when determining so-called *correspondence probabilities* (Section 5.2); it is not so important in the determination of likelihood ratios (Section 2.5.1). Reference to form (source or receptor, bulk or transferred particle) is a reference to one of the two parts of the evidence. Reference to anchoring (scene or suspect) is a reference to a perspective for the evaluation of the evidence.

Other terms such as control, known, recovered, questioned, will be avoided as far as possible. However, it is sometimes useful to refer to a sample of material found at the scene of a crime as the *crime sample* and to a sample of material found on or about a suspect as the *suspect sample*. This terminology reflects the site at which the material was found. It does not indicate the kind of material (bulk or transferred particle form) or the perspective (scene- or suspect-anchored) by which the evidence will be evaluated.

1.5   TYPES OF DATA

A generic name for observations which are made on items of interest, such as blood stains, refractive indices of glass, etc. is *data*. There are different types of data and some terminology is required to differentiate amongst them. In Example 1.1, the observations of interest are the blood groups of the crime stain and of the suspect. These are not quantifiable. There is no numerical significance which may be attached to these. The blood group is a qualitative characteristic. As such, it is an example of so-called *qualitative data*. The observation of interest is a quality, the blood group, which has no numerical significance. The different blood groups are sometime known as *categories*. The assignation of a person to a particular category is called a *classification*. A person may be said to be classified into one of several categories.

It is not possible to order blood groups and say that one is larger or smaller than another. However, there are other qualitative data which do have a natural ordering, such as the level of burns on a body. There is not a numerical measure of this but the level of burns may be classified as first, second, third degree, for example. Qualitative data which have no natural ordering are known as *nominal* data. Qualitative data to which a natural ordering may be attached are known as *ordinal* data. The simplest case of nominal data arises when an observation (e.g., of a person's blood group) may be classified into one of only two possible categories. For example, consider the Kell genetic marker system where a person may be classified as either Kell + or Kell −. Such data are known as *binary*. Alternatively, the variable of interest, here the Kell system, is known as *dichotomous*.

Other types of data are known as *quantitative* data. These may be either counts (known as *discrete* data, since the counts take discrete, integer values) or measurements (known as *continuous* data, since the measurements may take any value on a continuous interval).

A violent crime involving several people, victims and offenders, may result in much blood being spilt and many stains from each of several blood groups being identified. Then the numbers of stains in each of the different groups are examples of discrete, quantitative data.

The refractive indices of glass fragments and the positions of the bands on a DNA profile are both examples of continuous measurements. In practice, variables are rarely truly continuous because of the limits imposed by the sensitivity of the measuring instruments. Refractive indices or the band positions in a DNA profile may be measured only to a certain accuracy. (The DNA profile may be considered, perhaps, to be a pair of discrete variables since it corresponds to two numbers of base pairs. However, this is a situation in which the numbers have so many significant figures, typically four or five, that very little information is lost by treating the variable as continuous.)

Observations, or data, may thus be classified as qualitative or quantitative. Qualitative data may be classified further as nominal or ordinal and quantitative data may be classified further as discrete or continuous.

1.6 PROBABILITY

1.6.1 Introduction

The interpretation of scientific evidence may be thought of as the assessment of a comparison. This comparison is that between evidential material found at the scene of a crime (denote this by M_c) and evidential material found on a suspect, a suspect's clothing or around his environment (denote this by M_s). Denote the combination by $M = (M_c, M_s)$. As a first example, consider the blood stains of Example 1.1. The crime stain is M_c, the receptor or transferred particle form of the evidential material, M_s is the blood group of the suspect and is the source form of the material. From Example 1.2, suppose glass is broken during the commission of a crime. M_c would be the fragments of glass (the source form of the material) found at the crime scene, M_s would be fragments of glass (the receptor or transferred particle form of the material) found on the clothing of a suspect and M would be the two sets of fragments.

Qualities, such as the blood groups, or measurements, such as the refractive indices of the glass fragments are taken from M. Comparisons are made of the source form and the receptor form. Denote these by E_c and E_s, respectively, and let $E = (E_c, E_s)$ denote the combined set. Comparison of E_c and E_s is to be made and the assessment of this comparison has to be quantified. The totality of the evidence is denoted Ev and is such that $Ev = (M, E)$.

Statistics has developed as a subject, one of whose main concerns is the quantification of the assessments of comparisons. The performance of a new treatment, drug or fertiliser has to be compared with that of an old treatment, drug or fertiliser, for example. It seems natural that statistics and forensic science should come together. Two samples, the crime and the suspect sample, are to be compared. Yet, apart from the examples discussed in Chapter 4, it is only recently that this has happened. As discussed in Section 1.2 there have been several books describing the role of statistics in the law. There have been not any concerned with statistics and the evaluation of scientific evidence. Two factors may have been responsible for this.

First, there has been a lack of suitable data from relevant populations. There was a consequential lack of a baseline against which measures of typicality of any characteristics of interest may be determined. One exception is the reference data which have been available for many years on blood group frequencies among certain populations. Not only has it been possible to say that the suspect's blood group matched that of a stain found at the scene of a crime, but also that this group is only present in, say, 0.01% of the population. Such data have not, until recently, been available in other areas of forensic interest. Now, data collections exist for the refractive index of glass fragments (Lambert and Evett, 1984) and the complementary chromaticity coordinates measuring colour in fibres (Laing *et al.*, 1986; Laing *et al.*, 1987), to name but two. There is, also, much information about the frequency of characteristics in DNA profiles.

Secondly, the approach adopted by forensic scientists in the assessment of their evidence has been difficult to model; it has been one of comparison and significance. Characteristics of the crime and suspect samples are compared. If the examining scientists believe them to be similar, the typicality, and hence the significance of the similarity, of the characteristics is then assessed. This approach is what has been modelled by the two-stage approach of Evett (1977), described briefly in Section 1.3.2 and in fuller detail in Chapter 4. However, interpretation of the results provided by this approach is difficult.

Then, in a classic paper, Lindley (1977a) described an approach which was easy to justify, to implement and to interpret. It combined the two parts of the two-stage approach into one statistic and is discussed in detail in Section 7.2.4.

The approach compares two probabilities, the probability of the evidence, assuming one hypothesis about the suspect to be true (that he is guilty, for example) and the probability of the evidence, assuming another hypothesis about the suspect to be true (that he is innocent, for example). This approach implies that it is not enough for a prosecutor to show that evidence is unlikely if a suspect is innocent. The evidence has also to be more likely if the suspect is guilty. Such an approach had a good historical pedigree (Good, 1950, and see also Good, 1991, for a review) yet it had received very little attention in the forensic science literature. It is also capable of extension beyond the particular type of example discussed by Lindley, as will be seen by the discussion throughout this book, for example in Section 7.4.

However, in order to proceed it is necessary to have some idea about how uncertainty can be measured. This is best done through probability.

1.6.2   A standard for uncertainty

An excellent description of probability and its role in forensic science has been given by Lindley (1991). Lindley's description starts with the idea of a standard for uncertainty. He provides an analogy using the concept of balls in an urn. Initially, the balls are of two different colours, black and white. In all other respects, size, weight, texture etc., they are identical. In particular, if one were to pick a ball from the urn, without looking at its colour, it would not be possible to tell what colour it was. The two colours of balls are in the urn in proportions b and w for black and white balls, respectively, such that $b + w = 1$. For example, if there were 10 balls in the urn, of which 6 were black and 4 were white, then $b = 0.6$, $w = 0.4$ and $b + w = 0.6 + 0.4 = 1$.

The urn is shaken up and the balls thoroughly mixed. A ball is then drawn from the urn. Because of the shaking and mixing it is assumed that each ball, regardless of colour, is equally likely to be selected. Such a selection process, in which each ball is equally likely to be selected, is known as a random selection, and the chosen ball is said to have been chosen *at random*.

The ball, chosen at random, can be either black, an event which will be denoted B, or white, an event which will be denoted W. There are no other possibilities; one and only one of these two events has to occur. The uncertainty of the event B, the drawing of a black ball is related to the proportion b of black balls in the urn. If b is small (close to zero), B is unlikely. If b is large (close to 1), B is likely. A proportion b close to $1/2$ implies that B and W are about equally likely. The proportion b is referred to as the probability of obtaining a black ball on a single random drawing from the urn. In a similar way, the proportion w is referred to as the probability of obtaining a white ball on a single random drawing from the urn.

Notice that on this simple model, probability is represented by a proportion. As such it can vary between 0 and 1. A value of $b = 0$ occurs if there are no black balls in the urn and it is, therefore, impossible to draw a black ball from the urn.

The probability of obtaining a black ball on a single random drawing from the urn is zero. A value of $b = 1$ occurs if all the balls in the urn are black. It is certain that a ball drawn at random from the urn will be black. The probability of obtaining a black ball on a single random drawing from the urn is one. All values between these extremes of 0 and 1 are possible (by considering very large urns containing very large numbers of balls).

A ball has been drawn at random from the urn. What is the probability that the selected ball is black? The event B is the selection of a black ball. Each ball has an equal chance of being selected. The colours black and white of the balls are in the proportions b and w. The proportion, b, of black balls corresponds to the probability that a ball, drawn in the manner described (i.e. at random) from the urn is black. It is then said that the probability a black ball is drawn from the urn, when selection is made at random, is b. Some notation is needed to denote the probability of an event. The probability of B, the drawing of a black ball, is denoted $Pr(B)$ and similarly $Pr(W)$ denotes the probability of the drawing of a white ball. Then it can be written that $Pr(B) = b$, $Pr(W) = w$. Note that

$$Pr(B) + Pr(W) = b + w = 1.$$

This concept of balls in an urn can be used as a reference for considering uncertain events. Let R denote the uncertain event that the England football team will win the next European Football Championship. Let B denote the uncertain event that a black ball will be drawn from the urn. A choice has to be made between R and B and this choice has to be ethically neutral. If B is chosen and a black ball is indeed drawn from the urn then a prize is won. If R is chosen and England do win the Championship the same prize is won. The proportion b of black balls in the urn is known in advance. Obviously, if $b = 0$ then R is the better choice, assuming, of course, that England do have some non-zero chance of winning the Championship. If $b = 1$ then B is the better choice. Somewhere in the interval $[0, 1]$, there is a value of b, b_0 say, where the choice does not matter. One is indifferent as to whether R or B is chosen. If B is chosen, $Pr(B) = b_0$. Then it is said that $Pr(R) = b_0$, also. In this way the uncertainty in relation to any event can be measured by a probability b_0, where b_0 is the proportion of black balls which leads to indifference between the two choices, namely the choice of drawing a black ball from the urn and the choice of the uncertain event in whose probability one is interested.

Notice, though, that there is a difference between these two probabilities. By counting, the proportion of black balls in the urn can be determined precisely. Probabilities of other events such as the outcome of the toss of a coin or the roll of a die are also relatively straightforward to determine, based on assumed physical characteristics such as fair coins and fair dice. Let H denote the event that when a coin is tossed it lands head uppermost. Then, for a fair coin, in which the outcomes of a head H or a tail T at any one toss are equally likely, the probability that the coin comes down head uppermost is $1/2$. Let F denote the event that when a die is rolled it lands 4 uppermost. Then, for a fair die, in which the outcomes 1, 2, ... , 6 at any one roll are equally likely, the probability the die lands 4 uppermost is $1/6$.

Probabilities relating to the outcomes of sporting events such as football matches or championships or horse races, or to the outcome of a civil or criminal trial, are rather different in nature. It may be difficult to decide on a particular value for b_0. The value may change as evidence accumulates, such as the results of particular matches and the fitness or otherwise of particular players, or the fitness of horses, the identity of the jockey, the going of the race track, etc. Also, different people may attach different values to the probability of a particular event. These kinds of probability are sometimes known as *subjective* or *personal* probabilities; see De Groot (1970). Another term is *measure of belief* since the probability may be thought to provide a measure of one's belief in a particular event. Despite these difficulties the arguments concerning probability still hold. Given an uncertain event R, the probability of R, $Pr(R)$, is defined as the proportion of balls b_0 in the urn such that if one had to choose between B (the event that a black ball was chosen) where $Pr(B) = b_0$ and R then one would be indifferent to which one was chosen. There are difficulties but the point of importance is that a standard for probability exists. A use of probability as a measure of belief is described in Section 6.5 where it is used to represent *relevance*. The differences and similarities in the two kinds of probability discussed above and their ability to be combined has been referred to as a *duality* (Hacking, 1975).

1.6.3   Events

The outcome of the drawing of a ball from the urn was called an event. If the ball was black, the event was denoted B. If the ball was white, the event was denoted W. It was not certain which of the two events would happen: would the ball be black, event B, or white, event W? The degree of uncertainty of the event (B or W) was measured by the proportion of balls of the appropriate colour (B or W) in the urn and this proportion was called the probability of the event (B or W). In general, for an event R, $Pr(R)$ denotes the probability that R occurs.

Events can be events which may have happened (past events), events which may be relevant at the present time (present events) and events which may happen in the future (future events). There is uncertainty associated with each of these. In each case, a probability may be associated with the event.

- Past event: a crime is committed and a blood stain of a particular type is found at the crime scene. A suspect is found. The event of interest is that the suspect left the stain at the crime scene. Though the suspect either did or did not leave the stain the knowledge of it is uncertain and hence the event can have a probability associated with it.

- Present event: a person is selected. The event of interest is that he is of blood group O. Again, before the result of a blood test is available, this knowledge is uncertain.
- Future event: The event of interest is that it will rain tomorrow.

All of these events are uncertain and have probabilities associated with them. Notice, in particular, that even if an event has happened, there can still be uncertainty associated with it. The probability that the suspect left the stain at the crime scene requires consideration of many factors, including the possible location of the suspect at the crime scene and the properties of transfer of blood from a person to a site. With reference to the blood group, consideration has to be given to the proportion of people in some population of blood group O. Probabilistic predictions are common with weather forecasting. Thus, it may be said, for example, that the probability it will rain tomorrow is 0.8.

1.6.4   Laws of probability

There are several laws of probability which describe the values which probability may take and how probability may be combined. These laws are given here, first for events which are not conditional on any other information and then for events which are conditional on other information.

The first law of probability has already been suggested implicitly.

First law of probability   Probability can take any value between 0 and 1, inclusive, and only one of those values. Let R be any event. Then $0 \leqslant Pr(R) \leqslant 1$. For an event which is known to be impossible, the probability is zero. Thus if R is impossible, $Pr(R) = 0$. (This law is sometimes known as the *convexity rule*, Lindley, 1991.)

Consider the hypothetical example of the balls in the urn of which a proportion b are black and a proportion w white, with no other colours present, such that $b + w = 1$. Proportions lie between 0 and 1; hence $0 \leqslant b \leqslant 1, 0 \leqslant w \leqslant 1$. For any event R, $0 \leqslant Pr(R) \leqslant 1$. Consider B, the drawing of a black ball. If this event is impossible then there are no black balls in the urn and $b = 0$. This law is sometimes strengthened to say that a probability can *only* be 0 when the associated event is known to be impossible.

The first law concerns only one event. The next two laws, sometimes known as the second and third laws of probability, are concerned with combinations of events. Events combine in two ways. Let R and S be two events. One form of combination is to consider the event 'R and S', the event that occurs if and only if R and S both occur. This is known as the *conjunction* of R and S.

Consider the roll of a six-sided fair die, that is, a die in which each of the six sides is equally likely to land uppermost in any one roll of the die. Let R denote the throwing of an odd number. Let S denote the throwing of a number greater than 3 (i.e., a 4, 5 or 6). Then the event 'R and S' denotes the throwing of a 5.

Secondly, consider rolling two six-sided fair die. Let R denote the throwing of a six with the first die. Let S denote the throwing of a six with the second die. Then the event 'R and S' denotes the throwing of a double 6.

The second form of combination is to consider the event 'R or S', the event that R occurs if R or S (or both) occurs. This is known as the *disjunction* of R and S.

Consider again the roll of a single six-sided fair die. Let R, the throwing of an odd number (1, 3 or 5), and S, the throwing of a number greater than 3 (4, 5 or 6), be as before. Then 'R or S' denotes the throwing of any number other than a 2 (which is both even and less than 3).

Secondly, consider drawing a card from a well-shuffled pack of 52 playing cards, such that each card is equally likely to be drawn. Let R denote the event that the card drawn is a spade. Let S denote the event that the card drawn is a club. Then the event 'R or S' is the event that the card drawn is from a black suit.

The second law of probability concerns the disjunction 'R or S' of two events. Events are called *mutually exclusive* when the occurrence of one excludes the occurrence of the other. For such events, the conjunction 'R and S' is imposs'' le. Thus $Pr(R \text{ and } S) = 0$.

Second law of probability   If R and S are mutually exclusive events, the probability of their disjunction 'R or S' is equal to the sum of the probabilities of R and S. Thus, for mutually exclusive events,

$$Pr(R \text{ or } S) = Pr(R) + Pr(S). \tag{1.2}$$

Consider the drawing of a card from a well-shuffled pack of cards with R defined as the drawing of a spade and S the drawing of a club. Then $Pr(R) = 1/4$, $Pr(S) = 1/4$, $Pr(R \text{ and } S) = 0$ (a card may be a spade, a club, neither but not both). Thus, the probability that the card is drawn from a black suit, $Pr(R \text{ or } S)$ is $1/2$, which equals $Pr(R) + Pr(S)$.

Consider the earlier example, the rolling of a single six-sided fair die. Then the events R and S are not mutually exclusive. In the discussion of conjunction it was noted that the event 'R and S' denoted the throwing of a 5, an event with probability $1/6$. The general law, when $Pr(R \text{ and } S) \neq 0$, is

$$Pr(R \text{ or } S) = Pr(R) + Pr(S) - Pr(R \text{ and } S).$$

This rule can be easily verified in this case where $Pr(R) = 1/2$, $Pr(S) = 1/2$, $Pr(R \text{ and } S) = 1/6$, $Pr(R \text{ or } S) = 5/6$.

The third law of probability concerns the conjunction of two events. Initially, it will be asumed that the two events are what is known as *independent*. By independence, it is meant that knowledge of the occurrence of one of the two events does not alter the probability of occurrence of the other event. As a simple example of independence, consider the rolling of two six-sided fair dice, A and B say. The outcome of the throw of A does not affect the outcome of the throw of

B. If A lands '6' uppermost, this result does not alter the probability that B will land '6' uppermost. The same argument applies if one die is rolled two or more times. Outcomes of earlier throws do not affect the outcomes of later throws.

Third law of probability   Let R and S be two independent events. Then

$$Pr(R \text{ and } S) = Pr(R) \times Pr(S). \tag{1.3}$$

This relationship is sometimes used as the definition of independence. Thus, two events R and S are said to be independent if $Pr(R \text{ and } S) = Pr(R) \times Pr(S)$ There is symmetry in this definition. Event R is independent of S and S is independent of R. This law may be generalised to more than two events. Consider n events $S_1, S_2, \ldots, S_n$. If they are mutually independent then

$$Pr(S_1 \text{ and } S_2 \text{ and } \ldots \text{ and } S_n) = Pr(S_1) \times Pr(S_2) \times \ldots \times Pr(S_n) = \prod_{i=1}^{n} Pr(S_i).$$

1.6.5   Dependent events and background information

Not all events are independent. Consider, one roll of a fair die with R, the throwing of an odd number as before, and S, the throwing of a number greater than 3, as before. Then, $Pr(R) = 1/2$, $Pr(S) = 1/2$, $Pr(R) \times Pr(S) = 1/4$ but $Pr(R \text{ and } S) = Pr(\text{throwing a} 5) = 1/6$.

Events which are not independent are said to be *dependent*. The third law of probability for dependent events, was first presented by Thomas Bayes (1764, see also Barnard,1958, and Pearson and Kendall, 1970). It is the general law for the conjunction of events. The law for independent events is a special case. Before the general statement of the third law is made, some discussion of dependence is helpful.

It is useful to consider that a probability assessment depends on two things: the event R whose probability is being considered and the information I (known as background information) available when R is being considered. The probability $Pr(R|I)$ is referred to as a *conditional probability*, acknowledging that R is conditional or dependent on I. Note the use of the vertical bar |. Events listed to the left of it are events whose probability is of interest. Events listed to the right are events whose outcomes are known and which may affect the probability of the events listed to the left of the bar, the vertical bar having the meaning 'given' or 'conditional on'.

Consider a defendant in a trial who may or may not be guilty. Denote the event that he is guilty by G. The uncertainty associated with his guilt, the probability that he is guilty, may be denoted by $Pr(G)$. It is a subjective probability. The uncertainty will fluctuate during the course of a trial. It will fluctuate as evidence is presented. It depends on the evidence. Yet, neither the notation, $Pr(G)$, nor the language, the probability of guilt, makes mention of this dependence. The probability of guilt at any particular time depends on the knowledge (or information) available at that time. Denote this information by I. It is then possible to speak of the probability of guilt given, or conditional on, the information available at that time. This is written as $Pr(G|I)$. If additional evidence E is presented this then becomes, along with I, part of what is known. What is taken as known is then 'E and I', the conjunction of E and I. The revised probability of guilt is $Pr(G|E$ and $I)$.

All probabilities may be thought of as conditional probabilities. Personal experience informs judgements made about events. For example, judgement concerning the probability of rain the following day is conditioned on personal experiences of rain following days with similar weather patterns to the current one. Similarly, judgement concerning the value of evidence or the guilt of a suspect is conditional on many factors. These include other evidence at the trial but may also include a factor to account for the reliability of the evidence. There may be eyewitness evidence that the suspect was seen at the scene of the crime but this evidence may be felt to be unreliable. Its value will then be lessened.

The value of scientific evidence will be conditioned on the background data relevant to the type of evidence being assessed. Evidence concerning frequencies of different types of blood groups will be conditioned on information regarding ethnicity of the people concerned for the values of these frequencies. Evidence concerning distributions of the refractive indices of glass fragments will be conditioned on information regarding the type of glass from which the fragments have come. The existence of such conditioning events will not always be stated explicitly. However, they should not be forgotten. As stated above, all probabilities may be thought of as conditional probabilities. The first two laws of probability can be stated in the new notation, for events R, S and information I as:

First law

$$0 \leqslant Pr(R|I) \leqslant 1 \tag{1.4}$$

and $Pr(\text{not } I|I) = 0$. If I is known, the event 'not I' is impossible and thus $Pr(I|I) = 1$.

Second law

$$Pr(R \text{ or } S|I) = Pr(R|I) + Pr(S|I) - Pr(R \text{ and } S|I). \tag{1.5}$$

Third law for independent events
The third law, assuming R and S independent, is

$$Pr(R \text{ and } S|I) = Pr(R|I) \times Pr(S|I). \tag{1.6}$$

Notice that the event I appears as a conditioning event in *all* the probability expressions. The laws are the same as before but with this simple extension. The third law for dependent events is given later by (1.7).

As an example of the use of the ideas of independence, consider a diallelic system in genetics in which the alleles are denoted A and a, with $Pr(A) = p$, $Pr(a) = q$; $Pr(A) + Pr(a) = p + q = 1$. This gives rise to three genotypes that, assuming Hardy–Weinberg equilibrium to hold, are expected to have the following probabilities

- p^2 (homozygotes for allele A),
- $2pq$ (heterozygotes),
- q^2 (homozygotes for allele a).

The genotype probabilities are calculated by simply multiplying the two allele probabilities together on the assumption that the allele inherited from one's father is independent of the allele inherited from one's mother. The factor 2 arises in the heterozygous case because two cases must be considered, that in which allele A was contributed by the mother and allele a by the father, and *vice versa*. Both of these cases has probability pq because of the assumption of independence. The particular locus under consideration is said to be in Hardy–Weinberg equilibrium when the two parental alleles are independent.

These results are represented diagrammatically in Table 1.3.

Suppose now that events R and S are dependent; i.e., that knowledge that R has occurred affects the probability that S will occur, and *vice versa*. For example, let R be the outcome of a draw of a card from a well-shuffled pack of 52 playing cards. This card is not replaced in the pack so there are now only 51 cards in the pack. Let S be the draw of a card from this reduced pack of cards. Let R be the event '*an Ace is drawn*'. Thus $Pr(R) = 4/52 = 1/13$. (Note here the conditioning information I that the pack is well-shuffled, with its implication that each of the 52 cards is equally likely to be drawn has been omitted for simplicity of notation; explicit mention of I will be omitted in many cases, but its existence should never be forgotten.) Let S be the event '*an Ace is drawn*' also. Then $Pr(S|R)$ is the probability that an Ace was drawn at the second draw, given that an Ace was drawn at the first draw (and given everything else that is known, in particular that the first card was not replaced). There are 51 cards at the time of the second draw of which three are Aces. (Remember that an Ace was drawn the first time which is the

Table 1.3 Genotype probabilities, assuming Hardy–Weinberg equilibrium, for a diallelic system with allele probabilities p and q

Allele from mother	Allele from father	
	$A(p)$	$a(q)$
$A(p)$	p^2	pq
$a(q)$	pq	q^2

information contained in R.) Thus $Pr(S|R) = 3/51$. It is now possible to formulate the third law of probability for dependent events.

Third law of probability for dependent events

$$Pr(R \text{ and } S|I) = Pr(R|I) \times Pr(S|R \text{ and } I). \tag{1.7}$$

Thus in the example of the drawing of the Aces from the pack, the probability of drawing two Aces is

$$Pr(R \text{ and } S|I) = Pr(R|I) \times Pr(S|R \text{ and } I) = \frac{4}{52} \times \frac{3}{51}.$$

Note that if the first card had been replaced in the pack and the pack shuffled again, the two draws would have been independent and the probability of drawing two Aces would have been $4/52 \times 4/52$.

Example 1.4   Consider the genetic markers Kell and Duffy. For both of these markers a person may be either positive ($+$) or negative ($-$). 60% of the people in a relevant population are Kell$+$ and 70% are Duffy$+$. An individual is selected at random from this population; i.e., an individual is selected using a procedure such that each individual is equally likely to be selected.

Consider the Kell marker first in relation to the urn example. Let those people with Kell$+$ correspond to black balls and those with Kell$-$ correspond to white balls. By analogy with the urn example, the probability that a randomly selected individual (ball) is Kell$+$ (black) corresponds to the proportion of Kell$+$ people (black balls) in the population, namely 0.6. Let R be the event that a randomly selected individual is Kell$+$. Then $Pr(R) = 0.6$. Similarly, let S be the event that a randomly selected individual is Duffy$+$. Then $Pr(S) = 0.7$.

If Kell and Duffy markers were independent, it would be calculated that the probability that a person selected at random were both Kell$+$ and Duffy$+$, namely $Pr(R \text{ and } S)$, was given by

$$Pr(R \text{ and } S) = Pr(R) \times Pr(S) = 0.6 \times 0.7 = 0.42.$$

However, it may not necessarily be the case that 42% of the population is (Kell$+$, Duffy$+$). Nothing in the information currently available justifies the assumption of independence used to derive this result. It is perfectly feasible that, say, 34% of the population are both Kell$+$ and Duffy$+$; i.e. $Pr(R \text{ and } S) = 0.34$. In such a situation where $Pr(R \text{ and } S) \neq Pr(R) \times Pr(S)$ it can be said that the Kell and Duffy genetic markers are not independent.

The information about the Kell and Duffy genetic markers can be represented in a tabular form, known as a 2×2 (two-by-two) table as in Table 1.4, the rows of

Table 1.4 The proportion of people in a population which fall into the four possible categories of genetic markers

Kell	Duffy		Total
	+	−	
+	34	26	60
−	36	4	40
Total	70	30	100

which refer to Kell (positive or negative) and the columns of which refer to Duffy (positive or negative). It is possible to verify the third law of probability for dependent events (1.7) using this table. Of the 60% of the population who are Kell+ (R), 34/60 are Duffy+ (S). Thus $Pr(S|R) = 34/60 = 0.68$. Also, $Pr(R) = 60/100$ and

$$Pr(R \text{ and } S) = Pr(R) \times Pr(S|R) = \frac{60}{100} \times \frac{34}{60} = 0.34,$$

as may be derived from the table directly.

This example also illustrates the symmetry of the relationship between R and S as follows. Of the 70% of the population who are Duffy+ (S), 34/70 are Kell+ (R). Thus $Pr(R|S) = 34/70$. Also $Pr(S) = 70/100$ and

$$Pr(R \text{ and } S) = Pr(S) \times Pr(R|S) = \frac{70}{100} \times \frac{34}{70} = 0.34.$$

Thus, for dependent events, R and S, the third law of probability, (1.7) may be written as

$$Pr(R \text{ and } S) = Pr(S|R) \times Pr(R) = Pr(R|S) \times Pr(S), \qquad (1.8)$$

where the conditioning on I has been omitted. The result for independent events follows as a special case with $Pr(R|S) = Pr(R)$ and $Pr(S|R) = Pr(S)$.

1.6.6   Law of total probability

This is sometimes known as the extension of the conversation (Lindley, 1991). Events $S_1, S_2, \ldots, S_n$ are said to be *mutually exclusive and exhaustive* if one of them has to be true and only one of them can be true; they exhaust the possibilities and the occurrence of one excludes the possibility of any other. Alternatively, they are called a *partition*. The event $(S_1$ or ... or $S_n)$ formed from the conjunction of the individual events $S_1, \ldots, S_n$ is certain to happen since the events are exclusive. Thus, it has probability 1 and

$$Pr(S_1 \text{ or } \ldots \text{ or } S_n) = Pr(S_1) + \cdots + Pr(S_n) = 1,$$

a generalisation of the second law of probability, (1.2), for exclusive events. The *ABO* system of blood grouping provides an example: the four types O, A, B, and AB are mutually exclusive and exhaustive.

Consider $n = 2$. Let R be any other event. The events 'R and S_1' and 'R and S_2' are exclusive. They cannot both occur. The event "'R and S_1' or 'R and S_2'" is simply R. Let S_1 be male, S_2 be female, R be left-handed. Then

1. 'R and S_1' denotes a left-handed male,
2. 'R and S_2' denotes a left-handed female.

The event "'R and S_1' or 'R and S_2'" is the event that a person is a left-handed male or a left-handed female which implies that the person is left-handed (R). Thus,

$$Pr(R) = Pr(R \text{ and } S_1) + Pr(R \text{ and } S_2)$$
$$= Pr(R|S_1)Pr(S_1) + Pr(R|S_2)Pr(S_2).$$

The argument extends to any number of mutually exclusive and exhaustive events to give the law of total probability.

Law of total probability   If $S_1, S_2, \ldots, S_n$ are n mutually exclusive and exhaustive events,

$$Pr(R) = Pr(R|S_1)Pr(S_1) + \cdots + Pr(R|S_n)Pr(S_n). \tag{1.9}$$

An example for blood types and paternity cases is given by Lindley (1991). Consider two possible groups, S_1 (Rh$-$) and S_2 (Rh$+$) for the father, so here $n = 2$. Assume the relative frequencies of the two groups are p and $(1 - p)$, respectively. The child is Rh$-$ (event R) and the mother is also Rh$-$ (event M). The probability of interest is the probability a Rh$-$ mother will have a Rh$-$ child, in symbols $Pr(R|M)$. This probability is not easily derived directly but the derivation is fairly straightforward if the law of total probability is invoked to include the father.

$$Pr(R|M) = Pr(R|M \text{ and } S_1)Pr(S_1|M) + Pr(R|M \text{ and } S_2)Pr(S_2|M). \tag{1.10}$$

This is a generalisation of the law to include information M. If both parents are

Rh $-$, event (M and S_1) then the child is Rh $-$ with probability 1, so $Pr(R|M$ and $S_1) = 1$. If the father is Rh $+$ (the mother is still Rh $-$), event S_2, then $Pr(R|M$ and $S_2) = 1/2$. Assume that parents mate at random with respect to the rhesus quality. Then $Pr(S_1|M) = p$, the relative frequency of Rh $-$ in the population, independent of M. Similarly, $Pr(S_2|M) = 1 - p$, the relative frequency of Rh $+$ in the population. These probabilities can now be inserted in (1.10) to obtain

$$Pr(R|M) = 1(p) + \tfrac{1}{2}(1 + p) = (1 + 3p)/2,$$

for the probability that a Rh $-$ mother will have a Rh $-$ child. This result is not intuitively obvious, unless one considers the approach based on the law of total probability.

A further extension of this law to consider probabilities for combinations of genetic marker systems in a racially heterogeneous population has been given by Walsh and Buckleton (1988). Let C and D be two genetic marker systems. Let S_1 and S_2 be two mutually exclusive and exhaustive subpopulations such that a person from the population belongs to one and only one of S_1 and S_2. Let $Pr(S_1)$ and $Pr(S_2)$ be the probabilities that a person chosen at random from the population belongs to S_1 and to S_2, respectively. Then $Pr(S_1) + Pr(S_2) = 1$. Within each subpopulation C and D are independent so that the probability that an individual chosen at random from one of these subpopulations is of type CD is simply the product of the individual probabilities. Thus

$$Pr(CD|S_1) = Pr(C|S_1) \times Pr(D|S_1),$$
$$Pr(CD|S_2) = Pr(C|S_2) \times Pr(D|S_2).$$

However, such a so-called *conditional independence* result does not imply unconditional independence (i.e., that $Pr(CD) = Pr(C) \times Pr(D)$). The probability that an individual chosen at random from the population is CD, without regard to his subpopulation membership, may be written as follows

$$Pr(CD) = Pr(CDS_1) + Pr(CDS_2)$$
$$= Pr(CD|S_1) \times Pr(S_1) + Pr(CD|S_2) \times Pr(S_2)$$
$$= Pr(C|S_1) \times Pr(D|S_1) \times Pr(S_1) + Pr(C|S_2) \times Pr(D|S_2) \times Pr(S_2).$$

This is not equal to $Pr(C) \times Pr(D)$.

1.6.7 Updating of probabilities

Notice that the probability of guilt is a subjective probability, as mentioned before (Section 1.6.4). Its value will change as evidence accumulates. Also, different people will have different values for it. The following examples, adapted from similar ones in De Groot (1970), illustrate how probabilities may change with increasing information. The examples have several parts and each part has to be considered in turn without information from a later part.

Example 1.5
(a) Consider four events S_1, S_2, S_3 and S_4. Event S_1 is that the area of Lithuania is no more than $50\,000\,\mathrm{km}^2$, S_2 is the event that the area of Lithuania is greater than $50\,000\,\mathrm{km}^2$ but no more than $75\,000\,\mathrm{km}^2$, S_3 is the event that the area of Lithuania is greater than $75\,000\,\mathrm{km}^2$ but no more than $100\,000\,\mathrm{km}^2$, and S_4 is the event that the area of Lithuania is greater than $100\,000\,\mathrm{km}^2$. Evaluate the probabilities of each of these four events. Remember that these are four mutually exclusive events and that the four probabilities should add up to 1. Which do you consider the most probable and what probability do you assign to it? Which do you consider the least probable and what probability do you assign to it?
(b) Now, consider the information that Lithuania is the 25th largest country in Europe (excluding Russia). Use this information to reconsider your probabilities in part (a).
(c) Consider the information that Estonia, which is the 30th largest country in Europe, has an area of $45\,000\,\mathrm{km}^2$ and use it to reconsider your probabilities from the previous part.
(d) Consider the information that Austria, which is the 21st largest country in Europe has an area of $84\,000\,\mathrm{km}^2$ and use it to reconsider your probabilities from the previous part.
The area of Lithuania is given at the end of the Chapter.

Example 1.6
(a) Imagine you are on a jury. The trial is about to begin but no evidence has been led. Consider the two events: S_1, the defendant is guilty; S_2, the defendant is innocent. What are your probabilities for these two events?
(b) The defendant is a tall Caucasian male. An eyewitness says he saw a tall Caucasian male running from the scene of the crime. What are your probabilities now for S_1 and S_2?
(c) A bloodstain at the scene of the crime was identified as coming from the criminal. It is of blood group Γ, with proportion 2% in the local Caucasian population. What are your probabilities now for S_1 and S_2?
(d) A window was broken during the commission of the crime. Fragments of glass were found on the defendant's clothing of a similar refractive index to that of the crime window. What are your probabilities now for S_1 and S_2?
(e) The defendant works as a demolition worker near to the crime scene. Windows on the demolition site have refractive indices similar to the crime window. What are your probabilities now for S_1 and S_2?
This example is designed to mimic the presentation of evidence in a court case. Part (a) asks for a prior probability of guilt before the presentation of any evidence. It may be considered as a question concerning the understanding of the dictum

'innocent until proven guilty'. See Section 2.5.1 for further discussion of this with particular reference to the logical problem created if a prior probability of zero is assigned to the event that the suspect is guilty.

Part (b) involves two parts. First, the value of the similarity in physical characteristics between the defendant and the person running from the scene of the crime, assuming the eyewitness is reliable, has to be assessed. Secondly, the assumption that the eyewitness is reliable has to be assessed.

In part (c) it is necessary to check that the defendant is of group Γ. It is not stated that he is but if he is not he should never have been a defendant. Secondly, is the local Caucasian population the correct population? The evaluation of evidence of the form in (c) is discussed in Chapter 6.

The evaluation of refractive index measurements mentioned in (d) is discussed in Chapter 7. Variation both within and between windows has to be considered. Finally, how information about the defendant's lifestyle may be considered is discussed in Chapter 5.

It should be noted that the questions asked initially in Example 1.6 are questions which should be addressed by the judge and/or jury. The forensic scientist is concerned with the evaluation of his evidence, not with probabilities of guilt or innocence. These are the concern of the jury. The jury combines the evidence of the scientist with all other evidence and uses its judgement to reach a verdict. The theme of this book is the evaluation of evidence. Discussion of issues relating to guilt or otherwise of suspects will not be very detailed.

(As a tail piece to this Chapter, the area of Lithuania is 65 301 km².)

2

The Evaluation of Evidence

2.1 ODDS

2.1.1 Complementary events

There is a measure of uncertainty, known as *odds* which will be familiar to people who know about gambling. Book-makers quote odds in sporting events such as horse races or football matches. For example, a particular horse may be given odds of '6 to 1 against' it winning a race or a particular football team may be quoted at odds of '3 to 2 on' to win a match or, equivalently, '3 to 2 in favour' of winning the match. Odds are equivalent to probability. The above phrases can be related directly to probability statements about the probability of the horse winning its race or the football team winning its match.

First, an event, known as the negation, or *complement*, of another event has to be introduced and given some notation. Let R be an event. Then the negation or complement of R is the event which is true when R is false and false when R is true. It is denoted $\bar{R}$ and read as '*R*-bar'. The events R and $\bar{R}$ are known as *complementary events*.

2.1.2 Examples

1. A coin is tossed. Let R be the event it lands heads. Then $\bar{R}$ is the event that it lands tails. If the coin is fair, $Pr(R) = 1/2$, $Pr(\bar{R}) = 1/2$.
2. A die is rolled. Let R be the event that a six is face uppermost. Then $\bar{R}$ is the event that a 1, 2, 3, 4 or 5 is rolled. If the die is fair, $Pr(R) = 1/6$, $Pr(\bar{R}) = 5/6$.
3. A person is checked to see if he is Kell+ or Kell−. Let R be the event that he is Kell+. Then $\bar{R}$ is the event that he is Kell−. Using the data in Table 1.4, $Pr(R) = 0.6$, $Pr(\bar{R}) = 0.4$.
4. A person is charged with a crime. Let G be the event that he is guilty. Then $\bar{G}$ is the event that he is not guilty.

Notice that the event 'R or $\bar{R}$', formed from the disjunction of R and its complement $\bar{R}$, is certain. Thus it has probability 1. Also, since R and $\bar{R}$ are mutually exclusive,

$$1 = Pr(R \text{ or } \bar{R}) = Pr(R) + Pr(\bar{R})$$

and hence

$$Pr(\bar{R}) = 1 - Pr(R).$$

In general, for complementary events R and $\bar{R}$,

$$Pr(R) + Pr(\bar{R}) = 1, \qquad (2.1)$$

a result known as the *additivity rule*. It is now possible to define odds.

2.1.3   Definition

If an event R has probability $Pr(R)$ of accurring, the *odds* against R are

$$Pr(\bar{R})/Pr(R).$$

From (2.1), the odds against R are

$$\frac{1 - Pr(R)}{Pr(R)}.$$

The odds in favour of R are

$$\frac{Pr(R)}{1 - Pr(R)}.$$

Given a probability for an event it is possible to derive the odds against the event. Given a value for the odds against an event it is possible to determine the probability of that event occurring. Thus, if the horse has odds of 6 to 1 against it winning the race and R is the event that it wins the race, then

$$\frac{1 - Pr(R)}{Pr(R)} = 6,$$

where '6 to 1' is taken as the ratio 6/1 and written as 6. Then

$$1 - Pr(R) = 6 \times Pr(R)$$

$$1 = \{6 \times Pr(R)\} + Pr(R)$$

$$= 7 \times Pr(R).$$

Thus $Pr(R) = 1/7$.

The phrases 'odds on' and 'odds in favour of' are equivalent and are used as the reciprocal of 'odds against'. Consider the football team which is '3 to 2 on' to win its match. The phrase '3 to 2' is taken as the ratio 3/2 as this is odds *on*. The relationship between odds and probability is written as

$$\frac{Pr(R)}{1 - Pr(R)} = \frac{3}{2}.$$

Thus

$$2 \times Pr(R) = 3\{1 - Pr(R)\},$$

$$5 \times Pr(R) = 3,$$

$$Pr(R) = 3/5.$$

The general result may be derived as follows. Let O denote the odds against the occurrence of an event R. Then

$$O = \frac{1 - Pr(R)}{Pr(R)}$$

$$O \times Pr(R) = 1 - Pr(R)$$

$$(O + 1) \times Pr(R) = 1$$

$$Pr(R) = \frac{1}{O + 1}.$$

This can be verified directly for the horse whose odds were 6 to 1 against it winning (with $O = 6$). For the football team with odds of 3 to 2 on this can be taken as 2 to 3 against ($O = 2/3$) and the result follows. Odds equal to 1 are known as *evens*.

The concept of odds is an important one in the evaluation of evidence. Evidence is evaluated for its effect on the probability of a certain supposition about a suspect (before he comes to trial) or defendant (while a trial is in progress). This supposition may be that the suspect was present at the crime scene and it is this supposition which will be most discussed in this book. Initially, however, the discussion will be in terms of the effect of evidence on the probabilities of the guilt (G) and the innocence ($\bar{G}$) of a suspect. These are two complementary events. The ratio of the probabilities of these two events, $Pr(G)/Pr(\bar{G})$, is the odds against innocence or the odds in favour of guilt. Notice, also, that the events are that the suspect is truly guilty or truly innocent, not that he is judged to be guilty or innocent. The same principles concerning odds also apply for conditional probabilities. Given background information I, the ratio $Pr(G|I)/Pr(\bar{G}|I)$ is the odds in favour of guilt, given I. Much of this book will be concerned with the effect on the odds in favour of a supposition about the suspect of the evidence E under consideration.

2.2 BAYES' THEOREM

2.2.1   Statement of the theorem

Consider the last two parts of the third law of probability as given in (1.7), namely that for events R and S,

$$Pr(S|R) \times Pr(R) = Pr(R|S) \times Pr(S).$$

If $Pr(R) \neq 0$, it is possible to divide by $Pr(R)$ and obtain the following.

Bayes' Theorem

For two events, R and S,

$$Pr(S|R) = \frac{Pr(R|S) \times Pr(S)}{Pr(R)} \qquad (2.2)$$

assuming $Pr(R) \neq 0$.

2.2.2   Examples

A numerical verification of this result is available from Table 1.4 and Example 1.4. Let R denote Kell$+$, S denote Duffy$+$. Then, as before $Pr(R|S) = 34/70$, $Pr(S) = 70/100$, $Pr(R) = 60/100$ and

$$\frac{Pr(R|S) \times Pr(S)}{Pr(R)} = \frac{(34/70) \times (70/100)}{(60/100)} = \frac{34}{60} = Pr(S|R).$$

The importance of Bayes' Theorem is that it links $Pr(S)$ with $Pr(S|R)$. The uncertainty about S as given by $Pr(S)$ on the right-hand side of (2.2) is altered by the knowledge about R to give the uncertainty about S as given by $Pr(S|R)$ on the left hand side of (2.2). Note that the connection between $Pr(S)$ and $Pr(S|R)$ involves $Pr(R|S)$ and $Pr(R)$.

The distinction between $Pr(S|R)$ and $Pr(R|S)$ is very important and needs to be recognised. In $Pr(S|R)$, R is known or given, S is uncertain. (In Example 1.4 the individual is assumed to be Kell$+$, it is uncertain as to whether he is Duffy$+$ or Duffy$-$.) In $Pr(R|S)$, S is known or given, R is uncertain. (In Example 1.4 the individual is assumed to be Duffy$+$, it is uncertain as to whether he is Kell$+$ or Kell$-$.)

Two further examples will emphasise the difference between these two conditional probabilities. First, let S be the event 'I have two arms and two legs' and let R be the event 'I am a monkey'. Then $Pr(S|R) = 1$, whereas $Pr(R|S) \neq 1$. The first probability is equivalent to saying that 'If I am a monkey then I have two arms and two legs'. The second probability is equivalent to saying that 'If I have two arms and two legs, I need not be a monkey'.

The second example is from Lindley (1991). Consider the following two statements.

Example 2.1

1. The death rate amongst men is twice that amongst women.
2. In the deaths registered last month there were twice as many men as women.

Let M denote male, F denote female, so that M and F are complementary events ($M \equiv \bar{F}, F \equiv \bar{M}$). Let D denote the event of death. Then statements (1) and (2) may be written as

$$1.\ Pr(D|M) = 2Pr(D|F),$$

$$2.\ Pr(M|D) = 2Pr(F|D).$$

Notice, also that $Pr(M|D) + Pr(F|D) = 1$ since M and F are complementary events. Equation (2.1) generalises to include a conditioning event (D in this case). Thus from statement (2),

$$1 - Pr(F|D) = 2Pr(F|D)$$

and

$$Pr(F|D) = 1/3, \quad Pr(M|D) = 2/3.$$

It is not possible to make any similar inferences from (1). Table 2.1 illustrates the point numerically. There are 100 males of whom two died, and 100 females of whom one died. Thus $Pr(D|M) = 0.02, Pr(D|F) = 0.01$, satisfying (1). There were three deaths in total, of whom two were male and one female, satisfying (2).

Example 2.2   Another example to illustrate the difference between the two probability statements $Pr(S|R)$ and $Pr(R|S)$ has been provided by Darroch (1987). Consider a town in which a rape has been committed. There are 10 000 men of suitable age in the town of whom 200 work underground at a mine. Evidence is found at the crime scene from which it is determined that the criminal is one of the 200 mineworkers. Such evidence may be traces of minerals which

Table 2.1   Hypothetical results for deaths amongst a population

	Male	Female	Total
Dead	2	1	3
Alive	98	99	197
Total	100	100	200

could only have come from the mine. A suspect is identified and traces of minerals, similar to those found at the crime scene are found on some of his clothing. How might this evidence be assessed?

Denote the evidence by E: the event that 'mineral traces have been found on clothing of the suspect which are similar to mineral traces found at the crime scene'. Denote the hypothesis that the suspect is guilty by G and the hypothesis that he is innocent by $\bar{G}$ (these are complementary hypotheses: one and only one is true).

A hypothesis may be thought of in a similar way to an event, if subjective probabilities are considered. Events may be measurements of characteristics of interest, such as concentrations of certain minerals within the traces. There may be a well-specified model representing the randomness in such measurements. However, the guilt or innocence of the suspect is something about which there is no well-specified model but about which it is perfectly reasonable for an individual to represent with a probability his state of uncertainty about the truth or otherwise of the hypotheses.

(Note the use of the word *hypothesis*. This is a technical statistical word which refers to a supposition of interest: one talks about *hypothesis testing* when testing suppositions. For example, above, there are two suppositions: that a suspect is guilty or that he is innocent. Hypotheses are complementary in the same way as events are said to be complementary. One and only one can be true and together they exhaust all possibilities.)

Assume that all people working underground at the mine will have mineral traces similar to those found at the crime scene on some of their clothing. This assumption is open to question but the point about conditional probabilities will still be valid. The probability of finding the evidence on an innocent person may then be determined as follows. There are 9999 innocent men in the town of whom 199 work underground at the mine. These 199 men will, as a result of their work, have this evidence on their clothing, under the above assumption. Thus $Pr(E|\bar{G}) = 199/9999 \simeq 200/10\,000 = 0.02$, a small number. Does this imply that a man who is found to have the evidence on him is innocent with probability 0.02? Not at all. There are 200 men in the town with the evidence (E) on them of whom 199 are innocent ($\bar{G}$). Thus $Pr(\bar{G}|E) = 199/200 = 0.995$. The equation of $Pr(E|\bar{G})$ with the probability $Pr(\bar{G}|E)$ is known as the *fallacy of the transposed conditional* (Diaconis and Freedman, 1981) and is discussed in more detail later in Sections 2.3.1 and 2.5.4

2.3   ERRORS IN INTERPRETATION

2.3.1   Fallacy of the transposed conditional

Examples of this fallacy abound. Consider the following example, from Gaudette and Keeping (1974). The authors conducted a lengthy experiment in an attempt to determine the ability of hair samples to distinguish between different people.

Multiple comparisons were made of hairs from many different people. In one experiment, nine hairs, selected to be mutually dissimilar, from one source were compared, one by one, with a single hair from another source. It was estimated, from the results of many such comparisons, that the probability that in at least one of the nine comparisons the two hairs examined, from different sources, would be indistinguishable would be 1/4500. The authors concluded that 'it is estimated that if one human scalp hair found at the scene of a crime is indistinguishable from at least one of a group of about nine dissimilar hairs from a given source, the probability that it could have originated from another source is very small, about 1 in 4500' (Gaudette and Keeping, 1974, p.605).

Let R denote the event that 'one human scalp hair found at the scene of a crime is indistinguishable from at least one of a group of about nine dissimilar hairs from a given source'. Let S denote the event that 'the nine dissimilar hairs come from a different source from the single hair'. Then, the authors' experiment gives a value for $Pr(R|S)$ but the authors' summarising statement gives a value for $Pr(S|R)$.

Other examples of the fallacy of the transposed conditional are given by Thompson and Schumann (1987) who gave it the name of the *prosecutor's fallacy*. For example:

'There is a 10% chance that the defendant would have the crime blood type if he were innocent. Thus there is a 90% chance that he is guilty.

or

The blood test is highly relevant. The suspect has the same blood type as the attacker. This blood type is found in only 1% of the population so there is only a 1% chance that the blood found at the scene came from someone other than the suspect. Since there is a 1% chance that someone else committed the crime there is a 99% chance that the suspect is guilty.'
(Thompson and Schumann, 1987, p.177.)

In general, let E denote the evidence and $\bar{G}$ the hypothesis that a suspect is innocent. A value is determined for $Pr(E|\bar{G})$, the probability of the evidence if the suspect is innocent. The interpretation of the value calculated can cause considerable confusion. Two special cases of the fallacy of the transposed conditional in which $Pr(E|\bar{G})$ is confused with

1. the probability that the suspect is not the source of the evidence (*source probability error*);
2. the probability that the suspect is not guilty (*ultimate issue error*);

are discussed by Koehler (1993).

2.3.2 Source probability error

Let $\bar{C}$ be the hypothesis that the evidence E was not left by the suspect.

A scientist determines a value for $Pr(E|\bar{C})$ as 1 in 7 million in a case in which evidence E is that a DNA match has been found between blood from a murder victim and blood recovered from the clothing of a suspect. Consider the following possible statement of the value of the evidence (based on Wike v. State, 596 So 2nd 1020, Fla S. Ct. 1992, transcript, pp. 147–148, given in Koehler, 1993).

With probability 1 in 7 million it can be said that the blood on the clothing of ` the suspect could be that of someone other than the victim.

Oher possibilities are, where the figures quoted are for illustrative purposes only:

- the probability that the DNA that was found at the scene of the crime came from anyone else other than the suspect is 1 in 7 million. . . . ;
- the probability of this DNA profile occurring at random is 1 in 18 billion; thus the likelihood that the DNA belongs to someone other than the suspect is 1 in 18 billion . . . ;
- the probability of finding the evidence on an innocent person is 0.01% (1 in 10 000) thus the likelihood that the suspect is guilty is 99.99%. . . ;
- the trace evidence has the same DNA profile as the suspect, thus the trace evidence has been left by the suspect. . . ;

None of the conclusions in these statements is justified by the value given by $Pr(E|\bar{C})$. All give a mistaken probability for the source of the evidence.

2.3.3   Ultimate issue error

The source probability error may be extended to an error known as the ultimate issue error (Koehler, 1993). This extends the hypothesis that the suspect is the source of the evidence to the hypothesis that the suspect is guilty. The case of *People v. Collins* (see Chapter 4) is a particular example of this. Consider a case in which $Pr(E|\bar{C})$ is 1 in 5 million, say, where, as before, $\bar{C}$ is the hypothesis that the evidence E was not left by the suspect. The ultimate issue error would interpret this as a probability of 1 in 5 million that the suspect was innocent.

2.3.4   Defender's fallacy

As well as the fallacy of the transposed conditional there is also a *defender's fallacy* (Thompson and Schumann, 1987). Consider the following illustrative statement from a defence lawyer:

> 'The evidence for blood types has very little relevance for this case. Only 1% of the population has the rare blood type found at the crime scene and in the suspect. However, in a city, like this one in which the crime occurred, with a population of 200 000 people who may have committed the crime this blood type would be found in approximately 2000 people. The evidence merely shows that the suspect is one of

2000 people in the city who might have committed the crime. The blood test evidence has provided a probability of guilt of 1 in 2000. Such a small probability has little relevance for proving the suspect is guilty.'

Strictly speaking the defence lawyer is correct. However, before the evidence of the blood test was available, the suspect had a probability of only 1 in 200 000 of being guilty (not accounting for any other evidence which may have been presented). The effect of the blood test evidence is to increase this probability by a factor of 100. The evidence is 100 times more likely if the suspect is *guilty* than if he is *innocent*. This may be thought to show that the blood test evidence is quite compelling in support of a hypothesis of guilt. Of course, on its own, this evidence is unlikely to be sufficient for a verdict of guilty to be returned.

Two other errors discussed by Koehler (1993) are the probability (another match) error and the numerical conversion error.

2.3.5 Probability (another match) error

As in Example 1.1, a crime is committed. Evidence E of a blood stain of group Γ is found at the scene and identified as belonging to the criminal. A suspect is identified. Let $\bar{C}$ be the hypothesis that the evidence E was not left by the suspect. Suppose the frequency of blood of the type in the stain is γ (amongst the relevant population) and so $Pr(E|\bar{C}) = \gamma$. Then the probability that a person selected at random from the population does not match this blood type is $(1 - \gamma)$. Let N be the size of the population. Then, assuming independence amongst the members of the population with respect to E, the probability of no matches with the crime stain blood type amongst the N members is $(1 - \gamma)^N$ (a generalisation to N events of the third law of probability for independent events, (1.3)). The complement of no matches is at least one match. Hence, the probability of at least one match is $\theta = 1 - (1 - \gamma)^N$. Two numerical examples are given in Table 2.2. Let $N = 1$ million and take γ as a value which has been estimated from some larger population, assumed similar to the relevant population with regard to blood group frequencies; see Smith and Charrow (1975) and Finney (1977) for comments about such a *super population*. Thus it is possible for γ to be less than $1/N$. The two probabilities for θ in Table 2.2 are considerably larger than the corresponding values for γ. The probability (another match) error arises when these two probabilities γ and θ are equated. In other words, a small value of γ is taken to imply a small value for the probability that at least one other person has the same

Table 2.2 Probability θ of at least one match, given a frequency of the evidence of γ, in a population of size 1 million

γ:	1/1 million	1/10 million
θ:	0.632	0.095

matching evidence. The results in Table 2.2 illustrate why this implication is false.

2.3.6 Numerical conversion error

Let γ be the blood group frequency of the crime stain as in Section 2.3.5. A match between the crime stain and the blood group of a suspect has been made. Let n be the number of people who would have to be tested before there is another match. It may be thought that the significance of the value of γ can be measured by equating $1/\gamma$ with n. A small value of γ implies a large value of n. It is fairly straightforward to calculate n, given γ and given some value for the probability of another match occurring, $Pr(M)$ say. The numerical conversion error claims that n equals $1/\gamma$ but this is not so.

Suppose $\gamma = 0.01$. There is a probability of 0.01 that a randomly selected individual would match the evidence E. The numerical conversion error would claim that 100 people need to be tested before another match might be expected, but this is not the case. The error is exposed by consideration of $Pr(M)$. Suppose, initially, that $Pr(M)$ is taken to be equal to 0.5. A value of n greater than the value determined using a value of $Pr(M)$ of 0.5 would imply that a match is more likely to happen than not if n people were tested.

Let n be the number of people who are to be tested before the probability of finding a match is greater than 0.5. The probability that a randomly selected individual does not match the evidence is $(1 - \gamma)$. For n independent randomly selected individuals, the probability that none matches the evidence is $(1 - \gamma)^n$. The probability there is at least one match is thus $1 - (1 - \gamma)^n$. Thus, for a match to be more likely than not with n individuals, $1 - (1 - \gamma)^n$ has to be greater than 0.5 and so

$$(1 - \gamma)^n < 0.5.$$

Table 2.3 Evidence occurs with frequency γ. Smallest number ψ of people to be observed before a match with the evidence occurs with a given probability, $Pr(M) = 0.5, 0.9$; $\psi_5 = \log 0.5/\log(1 - \gamma)$, $\psi_9 = \log 0.1/\log(1 - \gamma)$, n_5 is the smallest integer greater than ψ_5, n_9 is the smallest integer greater than ψ_9

	$Pr(M) = 0.5$		$Pr(M) = 0.9$	
γ	ψ_5	n_5	ψ_9	n_9
0.1	6.6	7	21.9	22
0.01	69.0	69	229.1	230
0.001	692.8	693	2301.4	2302

Table 2.4   The probability, θ', of at least one match with the evidence, which occurs with frequency γ, when $n' = 1/\gamma$ people are tested

γ	n'	θ'
0.1	10	0.651
0.01	100	0.634
0.001	1000	0.632

This inequality may then be written as $n \log(1 - \gamma) < \log 0.5$. Thus, $n > \log 0.5 / \log (1 - \gamma) = \psi_5$, say (remembering that here $(1 - \gamma)$ is less than 1 and so its logarithm is negative). A similar argument shows that if $Pr(M)$ is taken to be greater than 0.9 then $n > \log 0.1 / \log (1 - \gamma) = \psi_9$, say. Values of ψ and n are given in Table 2.3 for $\gamma = 0.1$, 0.01 and 0.001 and for values of $Pr(M)$ equal to 0.5 and 0.9.

It is also worth noting that if $n' = 1/\gamma$ people were tested this does not make the probability of a match certain. If n' people are tested, the probability of at least one match is $\theta' = 1 - (1 - \gamma)^{n'}$; see Table 2.4 for examples. Notice that as $\gamma \to 0$, $\theta' \to 1 - e^{-1} = 0.632.\ldots$ Notice also, that $n_5 < n'$. Thus the numerical conversion error, based on $Pr(M) = 0.5$, exaggerates the number of people that need to be tested before a match may be expected. This exaggerates the probative strength of a match and favours the prosecution.

2.4   THE ODDS FORM OF BAYES' THEOREM

2.4.1   Likelihood ratio

Replace S by $\bar{S}$ in (2.2) and the equivalent version of Bayes' Theorem is

$$Pr(\bar{S}|R) = \frac{Pr(R|\bar{S})Pr(\bar{S})}{Pr(R)} \qquad (2.3)$$

$(Pr(R) \neq 0)$.

The first equation (2.2) divided by the second (2.3) gives the following.

The odds form of Bayes' theorem

$$\frac{Pr(S|R)}{Pr(\bar{S}|R)} = \frac{Pr(R|S)}{Pr(R|\bar{S})} \times \frac{Pr(S)}{Pr(\bar{S})}. \qquad (2.4)$$

The left hand side is the odds in favour of S, given R has occurred. The right hand side is the product of two terms,

$$\frac{Pr(R|S)}{Pr(R|\bar{S})} \text{ and } \frac{Pr(S)}{Pr(\bar{S})}.$$

The latter of these is the odds in favour of S, without any information about R. The former is a ratio of probabilities but it is not in the form of odds. The conditioning events, S and $\bar{S}$, are different in the numerator and denominator, whereas the event R, the probability of which is of interest, is the same.

In the odds form of Bayes' Theorem, given here, the odds in favour of S are changed on receipt of information R by multiplication by the ratio $\{Pr(R|S)/Pr(R|\bar{S})\}$. This ratio is important in the evaluation of evidence and is given the name *likelihood ratio* or *Bayes' factor*. It is the ratio of two probabilities, the probability of R when S is true and the probability of R when S is false. Thus, to consider the effect of R on the odds in favour of S, i.e., to change $Pr(S)/Pr(\bar{S})$ to $Pr(S|R)/Pr(\bar{S}|R)$, the former is multiplied by the likelihood ratio. The odds $Pr(S)/Pr(\bar{S})$ are known as the *prior odds* in favour of S; i.e., odds prior to receipt of R. The odds $Pr(S|R)/Pr(\bar{S}|R)$ are known as the *posterior odds* in favour of S; i.e., odds posterior to receipt of R. With similar terminology, $Pr(S)$ is known as the *prior probability* of S and $Pr(S|R)$ is known as the *posterior probability* of S. Notice that to calculate the change in the odds on S, it is probabilities of R that are needed. The difference between $Pr(R|S)$ and $Pr(S|R)$, as explained before (Section 2.3), is vital. Consider two examples.

1. $Pr(R|S)/Pr(R|\bar{S}) = 3$; the event R is three times as likely if S is true than if S is false. The prior odds in favour of S are multiplied by a factor of 3.
2. $Pr(R|S)/(Pr(R|\bar{S}) = 1/3$; the event R is three times as likely if S is false than if S is true. The prior odds in favour of S are reduced by a factor of 3.

When considering the effect of R on S it is necessary to consider both the probability of R when S is true and the probability of R when S is false. It is a frequent mistake (the fallacy of the transposed conditional, Section 2.3.1, again) to consider that an event R which is unlikely if $\bar{S}$ is true thus provides evidence in favour of S. For this to be so, it is required additionally that R is not so unlikely when S is true. The likelihood ratio is then greater than 1 and the posterior odds is greater than the prior odds.

Notice that the likelihood ratio is a ratio of probabilities. It is greater than zero (except when $Pr(R|S) = 0$ in which case it is zero also) but has no theoretical upper limit. Probabilities take values between 0 and 1, inclusive; the likelihood ratio takes values between 0 and ∞. It is not an odds, however. Odds is the ratio of the probabilities of two complementary events, perhaps conditioned on some other event. The likelihood ratio is the ratio of the probability of the same event conditioned upon two exclusive events, though they need not necessarily be complementary. Thus, $Pr(R|S)/Pr(R|\bar{S})$ is a likelihood ratio; $Pr(S|R)/Pr(\bar{S}|R)$ is an odds statistic.

Notice that (2.4) holds also if S and $\bar{S}$ are hypotheses rather than events. Hypotheses may be complementary, such as presence (C) or absence $(\bar{C})$ of a suspect from a crime scene, but need not necessarily be so (see, for example, Example 2.4 and Section 8.6.2). In general, the two hypotheses to be compared will be known as competing hypotheses. The odds $Pr(S)/Pr(\bar{S})$ in such circumstances should be explicitly stated as odds in favour of S, relative to $\bar{S}$. In the special case in which the hypotheses are mutually exclusive and exhaustive, they are complementary. The odds may then be stated as the odds in favour of S, where the relationship to the complementary hypothesis is implicit.

Example 2.3   This hypothetical example is taken from Walsh and Buckleton (1991). Consider three events.

- let X_A refer to Kell phenotype: $A = \text{Kell} +$, $\bar{A} = \text{Kell} -$.
- let X_B refer to Duffy phenotype: $B = \text{Duffy} +$, $\bar{B} = \text{Duffy} -$.
- let X_C refer to colour: $C = \text{pink}$, $\bar{C} = \text{blue}$.

The hypothetical example, represented by Table 2.5, gives frequency counts in an area of 100 people. These frequency counts can be transformed into relative frequencies by dividing throughout by 100. These relative frequencies will now be considered as probabilities for the purpose of illustration. In practice, relative frequencies from a sample will only provide estimates of probabilities; the effects of such considerations are not discussed here.

From Table 2.5, the following probabilities, among others, may be derived:

$$Pr(A) = Pr(\text{Kell} +) = 50/100 = 0.5.$$

$$Pr(\bar{B}) = Pr(\text{Duffy} -) = 50/100 = 0.5.$$

$$Pr(A) \times Pr(\bar{B}) = 0.5 \times 0.5 = 0.25.$$

However,

$$Pr(A\bar{B}) = Pr(\text{Kell} +, \text{Duffy} -) = 16/100 = 0.16.$$

Thus

$$Pr(A\bar{B}) \neq Pr(A) \times Pr(\bar{B}).$$

The Kell and Duffy phenotypes are not, therefore, independent.

As well as the above joint probability, conditional probabilities can also be determined. Fifty people are Kell$+$ (A). Of these, 16 are Duffy$-$ $(\bar{B})$. Thus, $Pr(\bar{B}|A) \times Pr(A) = 0.32 \times 0.5 = 0.16 = Pr(A\bar{B})$ which provides verification of the third law of probability (1.8). Similarly, $Pr(A|\bar{B}) = 16/50 = 0.32$. Equation (2.2) may be verified by noting that $Pr(\bar{B}) = 0.5$ and so,

$$Pr(A|\bar{B}) = \frac{Pr(\bar{B}|A) \times Pr(A)}{Pr(\bar{B})} = \frac{0.32 \times 0.5}{0.5} = 0.32.$$

Table 2.5 Frequency of Kell and Duffy types by colour in a hypothetical area. (Reproduced with permission of The Forensic Science Society)

Kell	Duffy		Total
	+	−	
Pink			
+	32	8	40
−	8	2	10
Total	40	10	50
Blue			
+	2	8	10
−	8	32	40
Total	10	40	50
Combined			
+	34	16	50
−	16	34	50
Total	50	50	100

Now, consider only 'pink' (C) people.

$$Pr(A|C) = Pr(\text{Kell}+ \,|\, \text{pink}) = 40/50 = 0.8.$$

$$Pr(B|CA) = Pr(\text{Duffy}+ \,|\, \text{pink and Kell}+) = 32/40 = 0.8.$$

Thus,

$$Pr(AB|C) = Pr(B|AC) \times Pr(A|C) = 0.8 \times 0.8 = 0.64 = 32/50$$

from (1.7).

The odds form of Bayes' Theorem,

$$\frac{Pr(C|A)}{Pr(\overline{C}|A)} = \frac{Pr(A|C) \times Pr(C)}{Pr(A|\overline{C}) \times Pr(\overline{C})}$$

$$= \frac{Pr(A|C)}{Pr(A|\overline{C})} \times \frac{Pr(C)}{Pr(\overline{C})},$$

which is of particular relevance to the evaluation of evidence may also be verified. From Table 2.5, when C = 'pink', $\overline{C}$ = 'blue', then $Pr(C) = Pr(\overline{C}) = 0.5$. For someone of Kell+ (A) phenotype, $Pr(A|C) = 40/50 = 0.8$, $Pr(A|\overline{C}) = 10/50 = 0.2$. Thus

$$\frac{Pr(A|C)}{Pr(A|\overline{C})} \times \frac{Pr(C)}{Pr(\overline{C})} = \frac{0.8 \times 0.5}{0.2 \times 0.5} = 4. \tag{2.5}$$

Also,

$$Pr(A|C) = 40/50,$$

$$Pr(\bar{C}|A) = 10/50.$$

The ratio of these two probabilities is $Pr(C|A)/Pr(\bar{C}|A) = 4$, equal to (2.5). The odds form of Bayes' Theorem has been verified numerically.

Consider Example 2.3 as an identification problem. Before any information about Kell types, say, is available, a person is equally likely to be 'pink' or 'blue'. The odds in favour of one colour or the other is evens. It is then discovered that the person is Kell +. The odds is now changed, using the procedure in the example, to being 4 to 1 on in favour of 'pink'. The probability that the person is 'pink' is $4/5 = 0.8$. The effect or value of the evidence has been to multiply the prior odds by a factor of 4. The value of the evidence is 4.

2.4.2   Logarithm of the likelihood ratio

Odds and the likelihood ratio take values between 0 and ∞. Logarithms of these statistics take values on $(-\infty, \infty)$. Also, the odds form of Bayes' Theorem involves a multiplicative relationship. If logarithms are taken, the relationship becomes an additive one:

$$\log \left\{ \frac{Pr(S|R)}{Pr(\bar{S}|R)} \right\} = \log \left\{ \frac{Pr(R|S)}{Pr(R|\bar{S})} \right\} + \log \left\{ \frac{Pr(S)}{Pr(\bar{S})} \right\}.$$

The idea of evaluating evidence by adding it to the logarithm of the prior odds is very much in keeping with the intuitive idea of weighing evidence in the scales of justice. The logarithm of the likelihood ratio has been given the name *the weight of evidence* (Good, 1950). A likelihood ratio with a value greater than 1, which leads to an increase in the odds in favour of S, has a positive weight. A likelihood ratio with a value less than 1, which leads to a decrease in the odds in favour of S, has a negative weight. A positive weight may be thought to tip the scales of justice one way, a negative weight may be thought to tip the scales of justice the other way. A likelihood ratio with a value equal to 1 leaves the odds in favour of S and the scales unchanged.

Example 2.4   Consider two hypotheses regarding a coin:

- S: the coin is double-headed; if S is true, the probability of tossing a head equals 1, the probability of tossing a tail equals 0.
- $\bar{S}$: the coin has a head and a tail and is fair: if $\bar{S}$ is true the probability of tossing a head equals the probability of tossing a tail and both equal $1/2$.

Notice that these are not complementary hypotheses.

The coin is tossed ten times and the outcome of any one toss is assumed independent of the others. The result R is ten heads. Then $Pr(R|S) = 1$, $Pr(R|\bar{S}) = (\frac{1}{2})^{10}$. The likelihood ratio is

$$\frac{Pr(R|S)}{Pr(R|\bar{S})} = \frac{1}{(1/2)^{10}} = 2^{10} = 1024.$$

The evidence is 1024 times more likely if the coin is double headed. The weight of the evidence is $10\log(2)$. Each toss which yields a head contributes a weight $\log(2)$ to the hypothesis S that the coin is double headed. Suppose, however, that the outcome of one toss is a tail (T). Then $Pr(T|S) = 0$, $Pr(T|\bar{S}) = 1/2$. The likelihood ratio $Pr(T|S)/Pr(T|\bar{S}) = 0$ and the posterior odds in favour of S relative to $\bar{S}$ equals 0. This is to be expected. A double-headed coin cannot produce a tail. If a tail is the outcome of a toss then the coin cannot be double-headed.

A brief history of the use of the weight of evidence is given by Good (1991). Units of measurement are associated with it. When the base of the logarithms is 10, Turing suggested that the unit should be called a *ban* and one tenth of this would be called a *deciban*, abbreviated to *db*.

2.5   THE VALUE OF EVIDENCE

2.5.1   Evaluation of forensic evidence

Consider the odds form of Bayes' Theorem in the forensic context of assessing the value of some evidence. The initial discussion is in the context of the guilt or otherwise of the suspect. This may be the case, for example in the context of Example 1.1, if all innocent explanations for the blood stain have been eliminated. Later, greater emphasis will be placed on hypotheses that the suspect was, or was not, present at the scene of the crime. Event S is replaced by a hypothesis G, that the suspect (or defendant if the case has come to trial) is truly guilty. Event $\bar{S}$ is replaced by hypothesis $\bar{G}$, that the suspect is truly innocent. Event R is replaced by event Ev, the evidence under consideration. This may be written as $(E, M) = (E_c, E_s, M_c, M_s)$, the type of evidence and observations of it as in Section 1.6.1. The odds form of Bayes' Theorem then enables the prior odds (i.e., prior to the presentation of Ev) in favour of guilt to be updated to posterior odds given Ev, the evidence under consideration. This is done by multiplying the prior odds by the likelihood ratio which, in this context, is the ratio of the probabilities of the evidence assuming guilt and assuming innocence of the suspect. With this notation, the odds form of Bayes' Theorem may be written as

$$\frac{Pr(G|Ev)}{Pr(\bar{G}|Ev)} = \frac{Pr(Ev|G)}{Pr(Ev|\bar{G})} \times \frac{Pr(G)}{Pr(\bar{G})}.$$

Explicit mention of the background information I is omitted in general from probability statements, for ease of notation. With the inclusion of I the odds form of Bayes' Theorem is

$$\frac{Pr(G|Ev, I)}{Pr(\overline{G}|Ev, I)} = \frac{Pr(Ev|G, I)}{Pr(Ev|\overline{G}, I)} \times \frac{Pr(G|I)}{Pr(\overline{G}|I)}.$$

Notice the important point that in the evaluation of the evidence Ev it is two probabilities which are necessary: the probability of the evidence if the suspect is guilty and the probability of the evidence if the suspect is innocent. For example, it is not sufficient to consider only the probability of the evidence if the suspect is innocent and to declare that a small value of this is indicative of guilt. The probability of the evidence if the suspect is guilty has also to be considered.

Similarly, it is not sufficient to consider only the probability of the evidence if the suspect is guilty and to declare that a high value of this is indicative of guilt. The probability of the evidence if the suspect is innocent has also to be considered. An example of this is the treatment of the evidence of a bite mark in the Biggar murder in 1967–68 (Harvey *et al.*, 1968). In that murder a bite mark was found on the breast of the victim, a young girl, which had certain characteristic marks, indicative of the conformation of the teeth of the person who had bitten her. A 17-year old boy was found with this conformation and became a suspect. Such evidence would help towards the calculation of $Pr(Ev|G)$. However, there was no information available about the incidence of this conformation among the general public. Examination was made of 342 boys of the age of the suspect. This enabled an estimate—albeit an intuitive one—of $Pr(Ev|\overline{G})$ to be obtained and to show that the particular conformation found on the suspect was not at all common.

Consider the likelihood ratio $Pr(Ev|G)/Pr(Ev|\overline{G})$ further where explicit mention of I has again been omitted. This equals

$$\frac{Pr(E|G, M)}{Pr(E|\overline{G}, M)} \times \frac{Pr(M|G)}{Pr(M|\overline{G})}.$$

The second ratio in this expression, $Pr(M|G)/Pr(M|\overline{G})$, concerns the type and quantity of evidential material found at the crime scene and on the suspect. It may be written as

$$\frac{Pr(M_s|M_c, G)}{Pr(M_s|M_c, \overline{G})} \times \frac{Pr(M_c|G)}{Pr(M_c|\overline{G})}.$$

The value of the second ratio in this expression may be taken to be 1. The type and quantity of material at the crime scene is independent of whether the suspect is the criminal or someone else is. The value of the first ratio, which concerns the evidential material found on the suspect given the evidential material found at the crime scene and the guilt or otherwise of the suspect, is a matter for subjective judgement and it is not proposed to consider its determination further here. Instead, consideration will be concentrated on

$$\frac{Pr(E|G, M)}{Pr(E|\overline{G}, M)}.$$

In particular, for notational convenience, M will be subsumed into I and omitted, for clarity of notation. Then,

$$\frac{Pr(M|G)}{Pr(M|\overline{G})} \times \frac{Pr(G)}{Pr(\overline{G})}$$

which equals

$$\frac{Pr(G|M)}{Pr(\overline{G}|M)}$$

will be written as

$$\frac{Pr(G)}{Pr(\overline{G})}.$$

Thus

$$\frac{Pr(G|Ev)}{Pr(\overline{G}|Ev)} = \frac{Pr(G|E, M)}{Pr(\overline{G}|E, M)}$$

will be written as

$$\frac{Pr(G|E)}{Pr(\overline{G}|E)}$$

and

$$\frac{Pr(Ev|G)}{Pr(Ev|\overline{G})} \times \frac{Pr(G)}{Pr(\overline{G})}$$

will be written as

$$\frac{Pr(E|G)}{Pr(E|\overline{G})} \times \frac{Pr(G)}{Pr(\overline{G})}.$$

The full result is then

$$\frac{Pr(G|E)}{Pr(\overline{G}|E)} = \frac{Pr(E|G)}{Pr(E|\overline{G})} \times \frac{Pr(G)}{Pr(\overline{G})}, \qquad (2.6)$$

or if I is included

$$\frac{Pr(G|E,\ I)}{Pr(\bar{G}|E,\ I)} = \frac{Pr(E|G,\ I)}{Pr(E|\bar{G},\ I)} \times \frac{Pr(G|I)}{Pr(\bar{G}|I)}. \tag{2.7}$$

The likelihood ratio is the ratio

$$\frac{Pr(G|E,\ I)/Pr(\bar{G}|E,\ I)}{Pr(G|I)/Pr(\bar{G}|I)} \tag{2.8}$$

of posterior odds to prior odds. It is the factor which converts the prior odds in favour of guilt to the posterior odds in favour of guilt. The representation in (2.7) also emphasises the dependence of the prior odds on background information. Previous evidence may be included here also, see, for example, Section 5.1.3.

It is often the case that another representation may be appropriate. Sometimes it may not be possible to consider the effect of the evidence on the guilt or innocence of the suspect. However, it may be possible to consider the effect of the evidence on the possibility that there was contact between the suspect and the crime scene. For example, a blood stain at the crime scene may be of the same type as that of the suspect. This, considered in isolation, would not necessarily be evidence to suggest that the suspect was guilty, only that the suspect was at the crime scene. Consider the following two complementary hypotheses:

1. C: the suspect was at the crime scene;
2. $\bar{C}$: the suspect was not at the crime scene.

The odds form of Bayes' Theorem is then

$$\frac{Pr(C|E,\ I)}{Pr(\bar{C}|E,\ I)} = \frac{Pr(E|C,\ I)}{Pr(E|\bar{C},\ I)} \times \frac{Pr(C|I)}{Pr(\bar{C}|I)}. \tag{2.9}$$

The likelihood ratio converts the prior odds in favour of C into the posterior odds in favour of C.

The likelihood ratio may be thought of as the *value* of the evidence. Evaluation of evidence, the theme of this book, will be taken to mean the determination of a value for the likelihood ratio. This value will be denoted V.

Definition Consider two competing hypotheses, C and $\bar{C}$ and background information I. The *value* V of the evidence E is given by

$$V = \frac{Pr(E|C,\ I)}{Pr(E|\bar{C},\ I)}, \tag{2.10}$$

the likelihood ratio which converts prior odds $Pr(C|I)/Pr(\bar{C}|I)$ in favour of

Table 2.6   Effect on prior odds in favour of C relative to $\bar{C}$ of evidence E with value V of 1000. Reference to background information I is omitted

Prior odds $Pr(C)/Pr(\bar{C})$	V	Posterior odds $Pr(C\|E)/Pr(\bar{C}\|E)$
1/10 000	1000	1/10
1/100	1000	10
1 (evens)	1000	1000
100	1000	100 000

C relative to $\bar{C}$ into posterior odds $Pr(C|E, I)/Pr(\bar{C}|E, I)$ in favour of C relative to $\bar{C}$. An illustration of the effect of evidence with a value V of 1000 on the odds in favour of C, relative to $\bar{C}$, is given in Table 2.6.

2.5.2   Summary of competing hypotheses

Initially, when determining the value of evidence, the two competing hypotheses were taken to be that the suspect is guilty and that the suspect is innocent. However, these are not the only ones possible as the discussion at the end of the previous section has shown.

Care has to be taken in a statistical discussion about the evaluation of evidence, as to the purpose of the analysis. Misunderstandings can arise. For example, it has been said that 'the statistician's objective is a final, composite probability of guilt' (Kind, 1994). This may be an intermediate objective of a statistician on a jury. A statistician in such a position would then have to make a decision as to whether to find the defendant guilty or not guilty. Determination of the guilt, or otherwise, of a defendant is the duty of a judge and/or jury. The objective of a statistician advising a scientist on the worth of the scientist's evidence is rather different. It is to assess the value of the evidence under two competing hypotheses. The evidence which is being assessed will often be transfer evidence. The hypotheses may be guilt or innocence. However, in many cases they will not be. An illustration of the effect of evidence on the odds in favour of a hypothesis C relative to a hypothesis $\bar{C}$ is given in Table 2.6. However, it has to be emphasised that the determination of the prior odds is also a vital part of the equation. That determination is part of the duty of the judge and/or jury.

Several suggestions for competing hypotheses, including those of guilt and innocence, are given below.

1. G: the suspect is guilty;
   $\bar{G}$: the suspect is innocent.
2. C: there was contact between the suspect and the crime scene;
   $\bar{C}$: there was no contact between the suspect and the crime scene.

Table 2.7 Phenotypic frequencies for a white Californian population, from Berry and Geisser, 1986

Phenotype (Γ)	O	A	B	AB
Phenotypic frequency (γ)	0.479	0.360	0.123	0.038

3. C: the crime sample came from a Caucasian;
$\bar{C}$: the crime sample came from an Afro-Caribbean (Evett *et al.*, 1992a).
4. F: the alleged father is the true father of the child;
$\bar{F}$: the alleged father is not the true father of the child (Evett *et al.*, 1989c).
5. C: the two crime samples came from the suspect and one other man;
$\bar{C}$: the two crime samples came from two other men (Evett *et al.*, 1991).
6. C: the suspect was the person who left the crime stain;
$\bar{C}$: the suspect was not the person who left the crime stain (Evett *et al.*, 1989b).

In general, the two hypotheses can be referred to as the prosecutor's and defender's hypothesis, respectively. The prosecutor's hypothesis is the one to be used for the determination of the probability in the numerator, the defender's hypothesis is the one to be used for the determination of the probability in the denominator.

Example 2.5 (Example 1.1 continued)   A crime has been committed. A blood stain is found at the scene of the crime. All innocent sources of the stain are eliminated and the criminal is determined as the source of the stain. The purpose of the investigation is to determine the identity of the criminal; it is at present unknown. The blood stain is of a group Γ with a frequency γ in the relevant population. A suspect has been identified with the same blood group as the crime stain. These two facts, the blood group of the crime stain (E_c) and the blood group of the suspect (E_s) together form the observations on the evidence $E\{=(E_c, E_s)\}$. The likelihood ratio can be evaluated as follows. If the suspect is guilty, the match between the two blood groups is certain and $Pr(E|G) = 1$. If the suspect is innocent, the match is coincidental. The criminal is of blood group Γ. The probability that the suspect would also have group Γ is γ, the frequency of Γ in the relevant population. Thus, $Pr(E|\bar{G}) = \gamma$. The likelihood ratio is $Pr(E|G)/Pr(E|\bar{G}) = 1/\gamma$. The odds in favour of G are multiplied by $1/\gamma$.

Consider the following numerical illustration of this, for the *ABO* system of bloodgrouping. The four phenotypes and their frequencies for a white Californian population (Berry and Geisser, 1986, p.357) are given in Table 2.7. The effect $(1/\gamma)$ on the odds in favour of G is given in Table 2.8 for each phenotype. Verbal interpretations may be given to these figures. For example, if the crime stain were of type O, it could be said that 'the evidence of the blood group of the crime stain matching the suspect's blood group is twice as likely if the suspect is guilty than if he is innocent'. If the crime stain were of type AB, it could be said that 'the evidence ... is twenty six times as likely if the suspect is guilty than if he is innocent'.

Table 2.8   Value of the evidence for each phenotype

Phenotype	(Γ)	O	A	B	AB
Value	$(1/\gamma)$	2.09	2.78	8.13	26.32

Table 2.9   Qualitative scale for reporting the value of the support of the evidence for C against $\bar{C}$

1	$< V \leqslant$	$10^{1/2}$	Slight increase in support
$10^{1/2}$	$< V \leqslant$	$10^{3/2}$	Increase in support
$10^{3/2}$	$< V \leqslant$	$10^{5/2}$	Great increase in support
$10^{5/2}$	$< V$		Very great increase in support

2.5.3   Qualitative scale for the value of the evidence

This quantitative value has been given a qualitative interpretation (Jeffreys, 1983; Evett, 1987a). Consider two competing hypotheses C and $\bar{C}$ and a value V for a piece of evidence. Then, the qualitative scale suggested by Evett (1987a) is given in Table 2.9. Interestingly, Fienberg (1989) provides a quote from a nineteenth century jurist, Jeremy Bentham, which appears to anticipate the Jeffreys–Evett scale, though perhaps in application to the strength of the belief in the hypothesis of guilt rather than in the strength of the evidence.

> 'The scale being understood to be composed of ten degrees—in the language applied by the French philosophers to thermometers, a decigrade scale—a man says, My persuasion is at 10 or 9 etc. affirmative, or at least 10 etc. negative...' (Bentham, 1827; quoted in Fienberg, 1989.)

Notice, also, what is not said in the verbal interpretation. Consider blood of type O again. It is not claimed that the evidence is such that the suspect is twice as likely to be guilty as he would have been if the evidence had not been presented. It is the evidence which is twice as likely, not the hypothesis of guilt. In the Jeffreys–Evett scale a value of 2 would be said to slightly increase support in favour of C against $\bar{C}$. The largest value for $1/\gamma$ in Example 2.5 is 26.32, a value which lies between $10^{1/2}$ and $10^{3/2}$ and, hence, would be said to increase the support for C against $\bar{C}$. Statements concerning the probability of guilt require knowledge of the prior odds in favour of guilt, something which is not part of the scientist's knowledge.

2.5.4   Explanation of transposed conditional and defender's fallacies

Using the odds form of Bayes' Theorem (2.6) insight into the fallacy of the transposed conditional (or prosecutor's fallacy) and the defender's fallacy can be gained.

Fallacy of the transposed conditional

As before, a crime has been committed. A stain of blood has been found at the scene which has been identified as coming from the criminal. A suspect is identified and his blood group is the same as the crime stain. Let E denote the evidence that the suspect's blood group is the same as that of the crime stain. Let G denote the hypothesis that the suspect is guilty and its complement $\bar{G}$ denote the hypothesis that the suspect is innocent.

Consider the following two statements:

• the blood group is found in only 1% of the population,
• there is a 99% chance that the suspect is guilty.

The second statement does not follow from the first without an unwarranted assumption about the prior odds in favour of guilt. The first statement that the blood group is found in only 1% of the population is taken to mean that the probability that a person selected at random from the population has the same blood group as the crime stain is 0.01. Thus $Pr(E|\bar{G}) = 0.01$. Also, $Pr(E|G) = 1$. The value of V is 100.

The second statement is taken to mean that the posterior probability in favour of guilt (posterior to the presentation of E) is 0.99; i.e., $Pr(G|E) = 0.99$. Thus $Pr(\bar{G}|E) = 0.01$ since G and $\bar{G}$ are complementary hypotheses. The posterior odds are then $0.99/0.01$ or 99, which is approximately equal to 100. However, V also equals 100. From (2.6), the prior odds is approximately equal to 1:

$$Pr(G) \simeq Pr(\bar{G}).$$

The second statement of the fallacy of the transposed conditional follows from the first only if $Pr(G) \simeq Pr(\bar{G})$. In other words, the suspect is just as likely to be guilty as innocent. This is not in accord with the dictum that a person is innocent until proven guilty. The prosecutor's conclusion, therefore, does not follow from the first statement unless this unwarranted assumption is made.

Defender's fallacy

Assume the likelihood ratio is 100, as in the previous section. Consider the relevant population to contain 200 000 people. The defence says that there are 2000 people with the same blood group as the defendant. The probability that the defendant is guilty is 1/2000 and thus the evidence has very little value in showing this particular person guilty. As before $Pr(E|G) = 1$, $Pr(E|\bar{G}) = 0.01$ and V equals 100. Also, $Pr(G|E) = 1/2000$ and so $Pr(\bar{G}|E) = 1999/2000$. The posterior odds in favour of G are

$$\frac{1/2000}{1999/2000} = \frac{1}{1999} \simeq \frac{1}{2000}.$$

The prior odds equals the ratio of the posterior odds to V:

$$\frac{Pr(G)}{Pr(\overline{G})} = \frac{Pr(G|E)}{Pr(\overline{G}|E)} \bigg/ V \simeq \left(\frac{1}{2000}\right) \bigg/ 100 = \frac{1}{200\,000}.$$

Thus, $Pr(G) = 1/200\,001$, $Pr(\overline{G}) = 200\,000/200\,0001$. The prior probability of guilt is $1/200\,001$. The denominator is the size of the relevant population (of innocent people) plus one for the criminal. The implication is that everybody is equally likely to be guilty. This does seem in accord with the dictum of innocent until proven guilty. Colloquially, it could be said that the defendant is just as likely to be guilty as anyone else. The defender's fallacy is not really a fallacy. It is misleading, though, to claim that the evidence has little relevance for proving the suspect is guilty. Evidence which increases the odds in favour of guilt from $1/200\,000$ to $1/2000$ is surely relevant. On the Jeffreys–Evett scale, the evidence greatly supports the hypothesis of guilt.

Notice that it is logically impossible to equate the dictum 'innocent until proven guilty' with a prior probability of guilt of zero, $Pr(G) = 0$. If $Pr(G) = 0$, then $Pr(G|E) = 0$, from (2.6) no matter how overwhelming the evidence, no matter how large the value of V. The probability of guilt may be exceedingly small, so long as it is not zero. So long as the prior probability of guilt is greater than zero, it will be possible, given sufficiently strong evidence, to produce a posterior probability of guilt sufficiently large to secure a conviction. It could be argued that the defendant is as likely to be as guilty as anyone else; his prior probability of guilt would then be the reciprocal of the size of the population defined by 'anyone else'.

2.5.5   The probability of guilt

Another legal dictum is that the jury is told to find the suspect 'guilty' if it is persuaded that the prosecution has shown 'beyond reasonable doubt' that the suspect committed the crime. However, the meaning of 'beyond reasonable doubt' is not clear. Lord Denning has said that 'there may be degrees of proof within that standard' (Eggleston, 1983). The more serious the crime, the greater the presumption of innocence. There is evidence that this is how people behave in the results of a questionnaire administered to judges, jurors and students of sociology (Simon and Mahan, 1971) and summarised in Table 2.10. This gives the value for the probability of guilt which would be taken as proof beyond reasonable doubt by those groups of people for various crimes. The variability amongst these figures is surprisingly small. For judges, the range is from 0.92 to 0.87 and for jurors from 0.86 to 0.74. These probabilities may be used to determine odds in favour of guilt. For example, for judges the odds in favour of guilt for a charge of murder to be proved 'beyond reasonable doubt' are 0.92 to 0.08 or approximately 12 to 1. An alternative interpretation is that 1 out of 13

Table 2.10 Probability of guilt required for proof beyond reasonable doubt (Simon and Mahan, 1971)

| Crime | Mean of persons surveyed | | |
	Judges	Jurors	Students
Murder	0.92	0.86	0.93
Forcible rape	0.91	0.75	0.89
Burglary	0.89	0.79	0.86
Assault	0.88	0.75	0.85
Petty larceny	0.87	0.74	0.82

people convicted of murder by that standard would be innocent. For a charge of petty larceny the odds are approximately 7 to 1. These odds (7 to 1 and 12 to 1) are remarkably close when the different levels of punishment are considered. The results need careful interpretation. For example, it is unlikely that judges in America in 1971 were convicting, wrongly, 1 out of 13 people convicted of murder. It is far more likely that the judges do not have a good intuitive feel for the meaning of probability figures.

Consider the following illustration of the effect of changes in probability at the upper end of the scale on the odds. Suppose that a probability value of 0.999 for the probability of guilt was thought sufficient to prove someone guilty 'beyond reasonable doubt' The odds in favour of guilt are then 999 to 1. A change in the probability of guilt to 0.99 reduces the odds to 99 to 1. A reduction by a factor of 10 in the odds has been brought about by a change in the probability of less than 0.01.

Another interpretation of 'beyond reasonable doubt' may be given by consideration of (2.8) and the interpretation of the likelihood ratio as the factor which converts the prior odds in favour of guilt to the posterior odds in favour of guilt. Suppose the prior odds for the dictum 'innocent until proven guilty' are taken to be 1/1000, though in practice they may be a lot smaller. Suppose the posterior odds for the dictum 'proof beyond reasonable doubt' are taken to be 1000 to 1, though in practice they may be a lot bigger. Then the likelihood ratio, the factor which converts the prior odds into the posterior odds, has to take the value 1000/(1/1000) which is 1000 000 (1 million). In other words, the evidence has to be 1 million times more likely if the suspect is guilty than if he is innocent. Such evidence is very compelling and may reasonably be said to provide very strong support for the hypothesis that the suspect is guilty.

Determination of the probability of guilt has been a long-standing problem. The problem has often been approached by consideration of a finite population such as may be found on an island, of size $(N + 1)$ say. A crime is committed and evidence of a characteristic (e.g., a blood stain of group Γ, with frequency γ amongst some population) is found at the scene of the crime. A suspect is found who possesses this characteristic. What is the probability he is guilty? Such a problem has been dubbed the *island problem* (Eggleston, 1983; Yellin, 1979; Lindley, 1987).

Various solutions have been proposed. These include:

- $\{1 - (1 - \gamma)^{N+1}\}/\{(N + 1)\gamma\}$ (Yellin, 1979);
- $1/(1 + N\gamma)$ (Lindley, in correspondence with Eggleston—see Eggleston, 1983).

Balding and Donnelly (1994) and Dawid (1994), in the context of discussions of wider issues, show that the second of these is theoretically correct in the mathematical constraints of the problem as specified. It is, though, of limited practical value.

2.6 SUMMARY

Odds have been defined and Bayes' Theorem, relating conditional probabilities, presented. Various possible errors in the interpretation of probabilistic measures of the value of evidence have been discussed. The likelihood ratio and the odds version of Bayes' Theorem have been defined. The role of the likelihood ratio as the value of the evidence has been defined as the factor which converts prior odds in favour of guilt into posterior odds. This has enabled the fallacies of the transposed conditional and the defender to be explained. The next chapter discusses how probability may be evaluated in certain situations and how it is distributed over the set of possible outcomes. Given these general probability distributions, it is then possible to determine procedures for evaluating evidence within certain general frameworks.

3

Variation

So far events and the probability of their occurrence have been discussed. These ideas may be extended to consider counts and measurements about which there may be some uncertainty or randomness. In certain fairly general circumstances the way in which probability is distributed over the possible numbers of counts or values for the measurements can be represented mathematically. Three such sets of circumstances, two for counts and one for measurements will be described in Sections 3.3 and 3.4.1. However, before this can be done, certain other concepts have to be introduced.

3.1 POPULATIONS

'Who is "random man"?' This is the title of a paper by Buckleton *et al.* (1991). In order to evaluate the denominator of the likelihood ratio in (2.10) it is necessary to have some idea of the variability or distribution of the evidence under consideration within some population. This population will be called the *relevant* population (and a more formal definition will be given later in Section 5.4) because it is the population which is deemed relevant to the evaluation of the evidence. Variability is important because if the suspect did not commit the crime and is, therefore, assumed innocent it is necessary to be able to determine the probability of associating the evidence with him. Surveys of populations are required in order to obtain this information. Three such surveys for glass fragments, are Pearson *et al.* (1971), Harrison *et al.* (1985) and McQuillan and Edgar (1992). There are many surveys on the distribution of blood group frequencies (e.g., Gaensslen *et al.*, 1987a, b, c). Surveys for other materials are discussed in Chapter 6.

Care has to be taken when deciding how to choose the relevant population. Buckleton *et al.* (1991) describe two situations and explain how the relevant population is different for each.

The first situation is one in which there is transfer from the criminal to the crime scene as in Example 1.1 and discussed in greater detail in Section 5.3.1. In this situation, the details of any suspect are irrelevant under $\bar{C}$, the hypothesis that the suspect was not present at the scene of the crime. Consider a bloodstain at the crime scene which, from the background information I, it is possible to assume is blood from the criminal. If the suspect was not present, then clearly some other person must have left the stain. There is no reason to confine attention to any one group of people. In particular, attention should not be confined only to any group (e.g., ethnic group) to which the suspect belongs. However, if there is some information which might cause one to reconsider the choice of population then that choice may be modified. Such information may come from an eyewitness, for example, who is able to provide information about the offender's ethnicity. This would then be part of the background information I. In general, though, information about blood group frequencies would be required from a survey which is representative of all possible offenders. For evidence of blood groups it is known that age and sex are not factors affecting a person's blood group but that ethnicity is. It is necessary to consider the racial composition of the population of possible offenders (not suspects). Normally, it is necessary to study a general population since there will be no information available to restrict the population of possible criminals to any particular ethnic group or groups.

The second situation considered by Buckleton *et al.* (1991) is possible transfer from the crime scene to the suspect or criminal and discussed further in Section 5.3.2. The details of the suspect are now relevant even assuming he was not present at the crime scene. Consider the situation where there is a deceased victim who has been stabbed numerous times. A suspect, with a history of violence, has been apprehended with a heavy bloodstain on his jacket which is not of his own blood. What is the evidential value in itself, and not considering the blood grouping evidence, of the existence of such a heavy bloodstain, not of the suspect's blood? The probability of such an event (the existence of a heavy bloodstain) if the suspect did not commit the crime needs to be considered.

The suspect may offer an alternative explanation. The jury can then assign a probability to the occurrence of the evidence, given the suspect's explanation. The two hypotheses to be considered would then be:

- C: the blood was transferred during the commission of the crime,
- $\bar{C}$: the suspect's explanation is true,

and the jury would assess the evidence of the existence of transfer under these two hypotheses. Evaluation of the evidence of the blood group frequencies would be additional to this. The two parts could then be combined using the technique described in Section 5.1.3.

In the absence of an explanation from the suspect, the forensic scientist could conduct a survey of persons as similar as possible to the suspect in whatever are the key features of his behaviour or lifestyle. The survey would be conducted with respect to the suspect since it is of interest to learn about the transfer of bloodstains for people of the suspect's background. In a particular case, it may be that a survey of people of a violent background is needed. Results may be adequately provided by Briggs (1978) in which 122 suspects who were largely vagrants, alcoholics and violent sexual deviants were studied. The nature and lifestyle of the suspect determined the type of population to survey. Buckleton *et al.* (1991) reported also the work of Fong and Inami (1986) in which clothing items from suspects, predominantly in offences against the person, were searched exhaustively for fibres that were subsequently grouped and identified.

The idea of a relevant population is a very important one. Consider the example of offender profiling, one which is not strictly speaking forensic science but which is still pertinent. Consider the application to rape cases. Suppose the profiler is asked to comment on the offender's lifestyle, such as age, marital status, existence and number of previous convictions, and so on, which the profiler may be able to do. However, it is important to know something about the distribution of these in some general population. The question arises here, as in Buckleton *et al.* (1991) described above, as to what is the relevant population. In rape cases, it may not necessarily be the entire male population of the local community. It could be argued that it might be the population of burglars, not so much because rapists are all burglars first but rather burglars are a larger group of people who commit crimes involving invasion of someone else's living space. Information from control groups is needed, regarding both the distribution of observed traits among the general non-offending population and the distribution of similar offences amongst those without the observed traits.

3.2 SAMPLES AND ESTIMATES

For any particular type of evidence the distribution of the characteristic of interest is important. This is so that it may be possible to determine the rarity or otherwise of any particular observation. For a blood grouping system, the relative frequencies of each of the groups within the system is important. For the refractive index of glass, the distribution of the refractive index measurements is important. In practice, these distributions are not known exactly. They are estimated. The relative frequencies of the *ABO* blood grouping system amongst the population of S. E. England exist but are not known. Instead, they are estimated from a sample (see, e.g., Table 1.1). In the case of Table 1.1 the number of people in the sample was 23 179. Similarly, the distribution of the refractive indices of glass exists but is not known. Instead it may be estimated from a sample (see Table 7.5, extracted from Lambert and Evett, 1984, in which the number of fragments was 2269). Table 7.5 refers to float glass from buildings. Other types of glass discussed by Lambert and Evett (1984) include, among others, 'not float glass' from buildings, glass from vehicles (divided into windows and others, such as headlamps and driving mirrors) and container glass. In each of these situations, a sample has been observed (e.g., blood groups of people or measurements of the refractive index of glass fragments). These samples are assumed to be representative, in some sense, of some population (e.g., all Caucasians in S. E. England, Stedman, 1985, or all float glass from buildings, Lambert and Evett, 1984).

A characteristic of interest from the pplation is known as a *parameter*. The corresponding characteristic from the corresponding sample is known as an *estimate*. For example, the proportion γ_{AB} of Caucasians in S. E. England with group AB blood is a parameter. The proportion of Caucasians with group AB blood in the sample of 23 179 people studied by Stedman (1985) is an estimate of γ_{AB}. Conventionally a ˆ symbol is used to denote an estimate. Thus $\hat{\gamma}_{AB}$ denotes an estimate of γ_{AB}. From Table 1.1, $\hat{\gamma}_{AB} = 0.031$.

It is hoped that an estimate will be a good estimate in some sense. Different samples from the same population may produce different estimates. The proportion of AB people in a second sample of 23 179 people from S. E. England may have produced a different number of AB people and hence a different value for $\hat{\gamma}_{AB}$. Different results from different samples do not mean that some are wrong and others are right. They merely indicate the natural variability in the distribution of blood groups amongst people.

An estimate may be considered good if it is *accurate* and *precise*. Accuracy may be thought of as a measure of closeness of an estimate to the true value of the parameter of interest. In the example above it is obviously desirable that $\hat{\gamma}_{AB}$ be close to γ_{AB}. Precision may be thought of as a measure of the variability of the estimates, whether or not they are close to the true value (Kendall and Buckland, 1982, pp. 3, 152). If different samples lead to very different estimates of the same characteristic then the variability is great and the estimates are not very precise. For example, if the variability in the estimation procedure is large then a second estimate, from a different sample, of γ_{AB} may produce an estimate very different from 0.031 ($\hat{\gamma}_{AB}$), though given the very large sample used by Stedman (1985) such an outcome is extremely unlikely.

The importance of allowing for variability is illustrated by the following hypothetical example from a medical context. The reaction times of two groups of people, group A and group B say, are measured. Both groups have the same median reaction time, 0.20 seconds, but group A's times vary from 0.10 to 0.30 s whereas group B's range from 0.15 to 0.25 s. Samples of equal numbers of people from each group are then given a drug designed to reduce reaction times. In both cases, the reaction times of the samples of people given the drug range from 0.11 to 0.14 s. For group A, this is within the range of previous knowledge and there is a little, but not very strong, evidence to suggest that the drug is effective in reducing reaction times. For group B, however, the result is outwith the range of previous knowledge and there is very strong evidence to suggest that the drug is effective. Note that both group A and group B had the same initial median reaction time. The drug produced the same range of reaction times in samples from both groups. The distinction in the interpretation of the results between the two groups arises because of the difference in the range, or variability, of the results for the two groups.

Later, in Section 3.4.1, it will be seen that when measurements are standardised, variation is accounted for by including a measure of the variation, known as the *standard deviation*, in the process.

Another notational convention, as well as the ^ notation, uses Roman letters for functions evaluated from measurements from samples and Greek letters for the corresponding parameters from populations. Thus a sample mean may be denoted $\bar{x}$ and the corresponding population mean μ. A sample standard deviation may be denoted s and the corresponding population standard deviation σ. The square of the standard deviation is known as the *variance*; a sample variance may be denoted s^2, and the corresponding population variance σ^2.

The concept of a *random variable* (or *random quantity* or *uncertain quantity*, Lindley, 1991) needs some explanation also. A *random variable*, in a rather circular definition, is something which varies at random. For example, the number of sixes in four rolls of a die varies randomly amongst the five numbers $\{0, 1, 2, 3, 4\}$ as the die is rolled for several sets of four rolls and the number of sixes is a random variable. Similarly, the refractive index of a fragment of glass varies over the set of all fragments of glass and is a random variable. The variation to be considered in the refractive index of glass is, however, of a more complicated structure than the number of sixes in four rolls of a die. There is variation in refractive index within a window and between windows. This requires parameters to measure two standard deviations, one for each type of variation, and this problem is discussed in greater detail in Chapter 7.

Notation is useful in the discussion of random variables. Rather than write out in long-hand phrases such as 'the number of sixes in four rolls of a die' or 'the refractive index of a fragment of glass' the phrases may be abbreviated to a single upper-case Roman letter. For example, let 'X' be short for 'the number of sixes in four rolls of a die'. It then makes sense to write mathematically $Pr(X = 3)$, which may be read as 'the probability that the number of sixes in four rolls of a die equals 3'. More generally still, the 3 may be replaced by a lower-case Roman letter to give $Pr(X = x)$, say, where x may then be substituted by one of the permissible values $\{0, 1, 2, 3, 4\}$ as required.

Similarly, X may be substituted for 'the refractive index of a fragment of glass' and the phrase 'the probability that the refractive index of a fragment of glass is less than 1.5185' may be written as $Pr(X < 1.5185)$, or more generally as $Pr(X < x)$ for a general value x of the refractive index. For reasons which are explained later (Section 3.4.1) it is not possible to evaluate $Pr(X = x)$ for a random variable representing a continuous measurement.

The mean of a random variable is the mean of the corresponding population variable. In the examples above, this would be the mean number of sixes in the conceptual population of all possible sets of four rolls of a die (and here note that the population need not necessarily exist except as a concept) or the mean refractive index of the population of all fragments of glass (again, effectively, a conceptual population). The mean of a random variable is given a special name, the *expectation*, and for a random variable, X say, it is denoted $E(X)$. Similarly, the variance of a random variable is the corresponding population variance. For a random variable X, it is denoted $Var(X)$.

The applications of these concepts are now discussed in the context of probability distributions for counts and for measurements.

3.3   COUNTS

3.3.1   Probabilities

Suppose a fair six-sided die is to be rolled four times. Let the event of interest be the number of occurrences of a six being uppermost; denote this by X. Then X can take one of five different integer values, 0, 1, 2, 3 or 4. Over a sequence of groups of four rolls of the die, X will vary *randomly* over this set of five integers. Outcomes of successive rolls are independent. For any one group of four rolls of the die, X takes a particular value, one of the integers $\{0, 1, 2, 3, 4\}$. Denote this particular value by x.

There is a formula which enables this probability to be evaluated fairly easily. Notice that in any one roll, the probability of throwing a six is 1/6, the probability of not throwing a six is 5/6, as these are complementary events. Then

$$Pr(X = x) = \binom{4}{x}\left(\frac{1}{6}\right)^{x}\left(\frac{5}{6}\right)^{4-x}, \qquad x = 0, 1, \ldots, 4;$$

an example of the binomial distribution (Section 3.3.3). The term $(1/6)^{x}$ corresponds to the x sixes, each occurring independently with probability 1/6. The term $(5/6)^{4-x}$ corresponds to the $(4 - x)$ non-sixes, each occurring independently with probability 5/6. The term $\binom{4}{x}$ is the *binomial coefficient*

$$\binom{4}{x} = \frac{4!}{x!(4-x)!},$$

where $x! = x(x - 1)(x - 2) \cdots 1$ and, conventionally, $0! = 1$. The binomial coefficient here is the number of ways in which x sixes and $(4 - x)$ non-sixes may be selected from four rolls, without attention being paid to the order in which the sixes occur.

Suppose $x = 1$, there is one six and three non-sixes, then

$$Pr(X = 1) = \binom{4}{1}\left(\frac{1}{6}\right)^{1}\left(\frac{5}{6}\right)^{3}.$$

Now

$$\binom{4}{1} = \frac{4!}{1!3!} = \frac{4 \times 3 \times 2 \times 1}{1 \times 3 \times 2 \times 1} = 4,$$

$$\left(\frac{1}{6}\right)^{1} = \frac{1}{6},$$

$$\left(\frac{5}{6}\right)^3 = \frac{125}{216},$$

$$Pr(X = 1) = 4 \times \frac{1}{6} \times \frac{125}{216} = 0.3858.$$

The probabilities for the five possible outcomes relating to the number of sixes in four rolls of the die are given in Table 3.1. Notice that the sum of the probabilities is 1 since the five possible outcomes 0, 1, 2, 3 and 4 are mutually exclusive and exhaustive.

3.3.2 Summary measures

It is possible to determine a value for the mean of the number of sixes in four rolls of the die; this is the expectation of the number of sixes in four rolls of the die. Consider 10 000 groups of four rolls of the die. The probabilities in Table 3.1 may be considered as the expected proportion of times in which each of 0, 1, 2, 3 and 4 sixes would occur. Thus it would be expected that on 4823 times there would be 0 sixes, 3858 times 1 six, 1157 times 2 sixes, 154 times 3 sixes and on 8 times there would be 4 sixes. The total number of sixes expected is thus

$$(0 \times 4823) + (1 \times 3858) + (2 \times 1157) + (3 \times 154) + (4 \times 8) = 6666.$$

In any one group of four rolls the expected number $E(X)$ of sixes is then $6666/10\,000 = 0.6666$. Notice that this is not an achievable number (0, 1, 2, 3 or 4) but is justified by the calculations. (In a similar way, an average family size of 2.4 children is not an achievable family size.) There is a formula for its calculation:

$$E(X) = 0 \times Pr(X = 0) + \cdots + 4 \times Pr(X = 4)$$

$$= \sum_{x=0}^{4} x \, Pr(X = x).$$

This can be further shortened by denoting $Pr(X = x)$ by p_x so that

$$E(X) = \sum_{x=0}^{4} x p_x.$$

Table 3.1 Probabilities for the number of sixes, X, in four rolls of a fair six-sided die

Number of sixes (x)	0	1	2	3	4	Total
$Pr(X = x)$	0.4823	0.3858	0.1157	0.0154	0.0008	1.0000

In general, for $(n+1)$ outcomes $\{0, 1, \ldots, n\}$ with associated probabilities $p_0, p_1, \ldots, p_n$,

$$E(X) = \sum_{x=0}^{n} xp_x.$$

The expectation is a well-known statistic. Not so well-known is the variance which measures the variability in a set of observations. The number of sixes which occurs in any group of four rolls of the die varies from group to group over the integers 0, 1, 2, 3 and 4.

Consider the square of the difference, $d(x)^2 = \{x - E(X)\}^2$ between an outcome x and the expectation. This squared difference is itself a random variable and as such has an expectation. The expectation of $d(x)^2$, $E\{d(X)^2\}$, for a set of $(n+1)$ outcomes $\{0, 1, \ldots, n\}$ with associated probabilities $p_0, p_1, \ldots, p_n$ is

$$\sum_{x=0}^{n} \{x - E(X)\}^2 p_x,$$

and is the *variance* of X. The square root of the variance is the *standard deviation*. Another, quicker, method of evaluation of the variance is to evaluate

$$Var(X) = \sum_{x=0}^{n} x^2 p_x - \left(\sum_{x=0}^{n} xp_x \right)^2.$$

The variance may be worked out for the example of the number of sixes in four rolls of the die as follows, where $E(X) = 0.6666$. The variance of X may then be calculated as

$$Var(X) = \sum_{x=0}^{4} \{x - E(X)\}^2 p_x = \sum_{x=0}^{4} \{d(x)\}^2 p_x = 0.5557.$$

'The quicker way is to evaluate

$$Var(X) = \sum_{x=0}^{4} x^2 p_x - \left(\sum_{x=0}^{4} xp_x \right)^2 = 1.0000 - 0.6666^2 = 0.5556.$$

The intermediate calculations are given in Table 3.2.

This example of four rolls of a fair six-sided die may be generalised. Consider each roll of the die as a *trial*, in a statistical sense. At each trial there will be one of only two outcomes, a six or a non-six (0, 1, 2, 3, 4, 5). Conventionally, in general terms, these may be known as *success* and *failure*. The trials are independent of each other. The probability of each of the outcomes is constant from trial to trial

Table 3.2   Intermediate calculations for the variance of the number of sixes, x, in four rolls of a fair six-sided die

x	0	1	2	3	4
$d(x)$	-0.6666	0.3334	1.3334	2.3334	3.3334
$\{d(x)\}^2$	0.4444	0.1112	1.7780	5.4448	11.1116
p_x	0.4823	0.3858	0.1157	0.0154	0.0008
x^2	0	1	4	9	16

(the probability of a six is 1/6 for each roll). Such a set of trials is known as a set of Bernoulli trials (after the Swiss mathematician, James Bernoulli, 1654–1705).

3.3.3   Binomial distribution

Let n denote the number of trials. Let X denote the number of successes. Let p denote the probability of a success in any individual trial and let $q \{ = (1 - p)\}$ denote the probability of a failure in any individual trial. Let the probability, $Pr(X = x)$, that X, the number of successes, equals x, be denoted by p_x; $x = 0$, $1, \ldots, n$. This probability is dependent on n and p and more correctly should be written as $Pr(X = x | n, p)$.

The situation described above is a very common one. The distribution of the probabilities (*probability distribution*) over the set of possible outcomes is known as the *binomial distribution*. The function which gives the formula for the probabilities $Pr(X = x)$ is known as a *probability function*. For the binomial distribution $Pr(X = x)$ is given by

$$Pr(X = x) = \binom{n}{x} p^x (1 - p)^{n-x} = \binom{n}{x} p^x q^{n-x}, \qquad x = 0, 1, \ldots, n, \quad (3.1)$$

where

$$\binom{n}{x} = \frac{n!}{x!(n - x)!} \qquad (3.2)$$

the binomial coefficient.

It can be shown that

$$E(X) = np, \qquad Var(X) = npq.$$

(Verification of these formulae can be made by reference to the numerical results in the example above.)

The binomial distribution is only one of many distributions applicable in different situations in which the event of interest is a count. One other is of particular interest in sampling and is known as the *hypergeometric distribution*.

3.3.4 Hypergeometric distribution—sampling from small populations

A different aspect of sampling arises when it is necessary to decide how many transferred particle items of the same generic type are needed to make a comparison with a source sample (Bates and Lambert, 1991).

The probabilities of selecting samples of particular numbers of matching and non-matching items from small populations of items comprising different numbers of matching and non-matching items, may be determined using a probability distribution known as the hypergeometric distribution. The probability function for the hypergeometric distribution is a function of three binomial coefficients. The probability function is given below and then some examples of its use follow.

Consider a group of N transferred particle items. This group is considered to be the population and N, the population size, is assumed small. Amongst these transferred particle items, m match the source items and $N - m$ do not, though this is not known by the investigating scientist, where the definition of a match will be specific to the particular problem (e.g. concerning glass fragments). A sample of size r ($< N$) items is taken from this group. The random variable of interest is X, the number of items in the sample which match the source items. Then it can be shown that

$$Pr(X = x) = \frac{\binom{m}{x}\binom{N-m}{r-x}}{\binom{N}{r}}. \tag{3.3}$$

Example 3.1   Suppose there are 20 transferred particle fragments of glass found on the clothing of a suspect of which 10 match the source and 10 do not. A sample of six of these 20 fragments is selected at random for examination, so that each of the 20 fragments has an equal probability of being selected. The probability that three are found to match the source can be determined. Here, $N = 20$, $m = 10$, $r = 6$. Then from (3.3)

$$Pr(X = 3) = \frac{\binom{10}{3}\binom{10}{3}}{\binom{20}{6}}.$$

Since

$$\binom{10}{3} = 120 \quad \text{and} \quad \binom{20}{6} = 38\,760,$$

it can be calculated that

$$Pr(X = 3) = 120 \times 120/38\,760 = 0.37.$$

Further examples are given in Bates and Lambert (1991). With $N = 10$, $m = 2$ and $r = 5$, the probability that $X = 0$ is $0.22 (> 1/5)$. Thus, with 10 transferred particle items of which two match the source (and eight do not), if a sample of five items is examined, there is a probability greater than one in five that no matches will be found. Conversely, in practice, what is known is that $N = 10$, $r = 5$ and $x = 0$. There is a reasonably large probability (0.22) that no matching items will be found in a sample of size five, when, in fact, there are two. A conclusion from such an examination that there were no matching items, and hence that this evidence did not associate the suspect with the crime scene, would be wrong.

With 50 transferred particle items, 10 of which match and 40 which do not ($N = 50$, $m = 10$, the same proportions as in the previous paragraph) and a sample of size $r = 5$, then $Pr(X = 0) = 0.31$. There is about a one in three probability that none of the five items selected randomly will match and a similar erroneous conclusion to the one above would be made.

Tables of the hypergeometric distribution have been published. See, for example, Lindley and Scott (1984) for values of the probability distribution for $N \leqslant 17$. For larger values of N with r relatively small compared to N (e.g., $r < N/20$), probabilities may be calculated using the binomial distribution (3.1) with $p = m/N$, $n = r$. For example, with $N = 50$, $m = 10$, $r = 5$, $p = 0.2$ then

$$Pr(X = 0) = \binom{5}{0} 0.2^0 0.8^5 = 0.33,$$

which should be compared with an exact probability of 0.31. Notice that $r > N/20$ so that the approximation is not too good.

Suppose that $N = 120$, $m = 80$, $r = 5$, $p = 2/3$. The hypergeometric probability

$$Pr(X = 2) = \frac{\binom{80}{2}\binom{40}{3}}{\binom{120}{5}} = 0.164$$

The approximation from the binomial distribution is

$$Pr(X = 2) = \binom{5}{2}\left(\frac{2}{3}\right)^2\left(1 - \frac{2}{3}\right)^3 = 0.165,$$

and a very good approximation is obtained.

3.4   MEASUREMENTS

3.4.1   Normal distribution

When considering data in the form of counts, the variation in the possible outcomes can be represented by a *probability function*. The variation in measurements, which are continuous, may also be represented mathematically by a function, known as a *probability density function*. Probability functions and probability density functions are both examples of *probability models*.

As an example of a probability model for a continuous measurement, consider a case in which the evidence consists of DNA profiles, from a single-locus probe, of samples from a crime scene and from a suspect. Normally, they will be two-banded profiles but for the purposes of this example, single bands only are considered. Let the measurements be denoted y and x, respectively. Let $d = y - x$, the difference in band weights. Assume that the DNA profile in the crime sample did in fact come from the suspect. Under this assumption, d provides an estimate of measurement error. The variation in measurement error is represented by the probability density function which in this case is such that the function takes its greatest value when $d = 0$. As the difference between y and x increases, the value of the function decreases. The particular function which is used here is the Normal or Gaussian probability density function (named after the German mathematician Carl Friedrich Gauss, 1777–1855). There are other functions which could be used, e.g., the lognormal distribution (Section 8.5), but the Normal probability density function performs well enough for practical use (Evett *et al.*, 1992b).

The binomial distribution required the number of trials and the probability of a success to be known in order that the probability function could be defined. Two characteristics (or parameters) of the measurement error are required to define the Normal probability density function. These are the mean or expectation, θ, and the standard deviation, σ. The mean may be thought of as a *measure of location* to indicate the size of the measurements. The standard deviation may be thought of as a *measure of dispersion* to indicate the variability in the measurements. The square of the standard deviation, the variance, is denoted σ^2. Given these parameters, the Normal probability density function for d, $f(d|\theta, \sigma^2)$, is given by

$$f(d|\theta, \sigma^2) = \frac{1}{\sqrt{2\pi\sigma^2}}\exp\left\{-\frac{(d - \theta)^2}{2\sigma^2}\right\}.$$

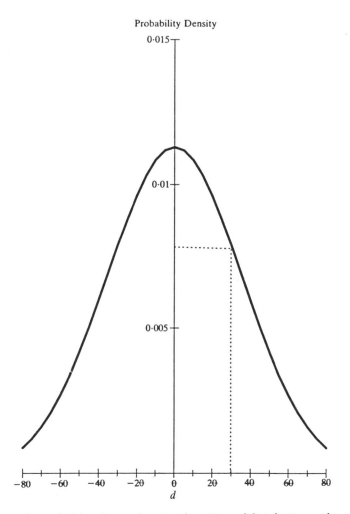

Figure 3.1   The probability density function for a Normal distribution with mean 0 and standard deviation $25\sqrt{2}$, representing the distribution of the difference d between measurements of band weights in *bp* on the same band in two separate profiles of the same DNA. The dotted line indicates the value 0.0079 of the probability density function when $d = 30\ bp$ (Evett *et al*, 1992b). (Reproduced with permission of The Forensic Science Society.)

where $\exp\{\cdots\}$ denotes e, the base of Napierian logarithms (e $= 2.718\ 281\ 828 \cdots$). The function $f(d|\theta, \sigma^2)$ is a function which is symmetric about θ. It takes its maximum value when $d = \theta$, it is defined on the whole real line for $-\infty < d < \infty$ and is always positive. The area underneath the function is 1, since d has to lie between $-\infty$ and ∞. It is illustrated in Figure 3.1 (Evett *et al.*, 1992b).

For the single-band DNA profile, if the crime and suspect bands come from the same person, the mean θ is 0. In this special case of zero mean, the Normal probability density function is then

$$f(d|0, \sigma^2) = \frac{1}{\sqrt{2\pi\sigma^2}} \exp\left(-\frac{d^2}{2\sigma^2}\right). \tag{3.4}$$

The Normal probability density function is so common that it has special notation. If the crime and suspect bands come from the same person, the measurement error d is Normally distributed with mean 0 and variance σ^2. This is abbreviated to

$$D \sim N(0, \sigma^2),$$

where $\sim$ is to be read as 'is distributed as', D is the random variable corresponding to d and the conditioning on $\theta = 0$ and σ^2 has been omitted for clarity. In general, a Normally distributed measurement, X say, with mean θ and variance σ^2, may be said to be such that

$$(X|\theta, \sigma^2) \sim N(\theta, \sigma^2).$$

The first symbol within parentheses conventionally denotes the mean, the second conventionally denotes the variance. It is not always necessary for the notation to make explicit the dependence of X on θ and σ^2. The distributional statement may then be denoted

$$X \sim N(\theta, \sigma^2),$$

and such abbreviated notation will be used often.

The determination of probabilities associated with Normally distributed random variables is made possible by a process known as *standardisation*, whereby a general Normally distributed random variable is transformed into one which has a *standard* Normal distribution, that is, one with zero mean and unit variance. Let

$$Z = (X - \theta)/\sigma.$$

Then, $E(Z) = 0$ and $Var(Z) = 1$. The random variable Z is Normally distributed with mean 0 and variance 1. This is a standard Normal distribution and the distribution of Z is denoted

$$Z \sim N(0, 1).$$

Standardisation is the process of transforming a Normal random variable X into a standard Normal random variable by subtracting the mean and dividing by the standard deviation. Notice that this process requires variability, as represented by σ to be taken into account. For example, the division by σ ensures the resulting statistic is dimensionless. If DNA profiles were being considered the statistic would be independent of whether fragment lengths were measured in base pairs or kilobase pairs.

Consider the following numerical example from Evett *et al.* (1992b). Let $x = 3000bp$. The standard deviation of the measurement error, σ_0 say, in this region of measurements has been shown in previous experiments to be $25\ bp$. Consider the measurements from the crime and suspect samples and assume these are independent. This is reasonable even if the two samples came from the same person, since it is the measurements that are of interest, not the people, or person, from which they came. The variance of the *difference* of two independent measurements is the *sum* of the variances of the two individual measurements. Thus the variance of the difference of these two (independent) measurements is $2\sigma_0^2$. The standard deviation of the difference is $\sigma_0\sqrt{2}$. In this example, the standard deviation of d is $25\sqrt{2}$. Suppose $y = 3{,}030\ bp$. Then $d = 30\ bp$. Substitution of $d = 30\ bp$, $\sigma^2 = 2\sigma_0^2 = 2 \times 25^2 = 1250$ into (3.4) gives

$$f(d) = f(30) = \frac{1}{\sqrt{2500\pi}}\exp\left(-\frac{900}{2500}\right) = 0.0079; \qquad (3.5)$$

see Figure 3.1.

Consider the continuous case in more detail. The function modelling the variation is known as a probability density function, not a probability function as it does not measure probabilities. An intuitive understanding of the terminology can be gained by considering the following analogy. A cylindrical rod, with circular cross-section has a density which varies along its length according to some function, f say. Then its weight over any particular part of its length is the integral of this function f over that part. In the same way, with a probability density function, the probability of a random variable lying in a certain interval is the integral of the corresponding density function over the interval. Thus the probability of the measurement error d lying within a certain interval would be the integral of $f(d)$ over this interval. Note, however, the following theoretical detail. A cross-section of zero thickness of the rod would have zero weight since its volume would be zero. Similarly, the probability of a continuous random variable taking a particular value is zero. In practice, measuring instruments are not sufficiently accurate to measure to an infinite number of decimal places and this problem does not arise so long as one determines the probability of a measurement lying within a particular interval and does not attempt to calculate the probability of a measurement taking a particular value. (See Section 4.4.5 for an application of this idea.)

3.4.2   Correlated measurements

Often, more than one characteristic is of interest, e.g., the refractive index and density of window glass. The data (measurements of these characteristics) are referred to as *multivariate data* and in the special case where only two characteristics are measured they are known as *bivariate data*. Let the variables be denoted $\mathbf{X} = (X_1, X_2, \ldots, X_p)'$, where the bold face $\mathbf{X}$ is used to denote a multivariate, or vector, set of observations and the $'$ is used to denote a column vector. For bivariate data $p = 2$. In the example of window glass, X_1 would be the refractive index, X_2, the density. For continuous data, the vector $\mathbf{X}$ has a probability density function, just as the individual characteristics have.

If the characteristics are independent then the joint probability density function $f(\mathbf{x})$ is the product of the individual probability density functions. Thus

$$f(\mathbf{x}) = f(x_1, \ldots, x_p) = \prod_{i=1}^{p} f(x_i)$$

which may be thought of as an extension of the third law of probability for independent events (1.3).

If the characteristics are not independent, however, such an approach is not possible. Assume the measurements of these characteristics are Normally distributed and dependent. The measurements are said to be *correlated*. The multivariate analogue of the Normal distribution may be obtained. The multivariate mean $\boldsymbol{\theta}$ is the vector formed by the means of the individual variables. Instead of a variance σ^2 there is a square $(p \times p)$ symmetric matrix Σ of variances and *covariances*. The covariance is a measure of the association between a pair of characteristics and is the product of the individual standard deviations and a factor which measures the *correlation* (degree of association) between the two characteristics. The variances of the p variables are located on the diagonal of Σ. The covariances are the off-diagonal terms so that the (i, j)-th cell of Σ contains the covariance between X_i and X_j (the covariance of X_i and X_i is simply the variance of X_i). The correlation between two variables is a parameter which measures the amount of linear association between the variables. It takes values between 0 and 1. Two variables which have a perfect linear relationship with a positive slope (as one increases so does the other) have a correlation of 1. Two variables which have a perfect linear relationship with a negative slope (as one increases, the other decreases) have a correlation of -1. A correlation of 0 implies that there is no linear association between the two variables. Notice that this does not mean there is no association between the variables, only that the association is not linear.

Let $Var(X_i) = \sigma_i^2$, $(i = 1, \ldots, p)$ and let ρ_{ij} denote the correlation between X_i and X_j, $(i = 1, \ldots, p; j = 1 \ldots, p; i \neq j)$. Then the covariance between X_i and X_j, denoted $Cov(X_i, X_j)$ is given by

$$Cov(X_i, X_j) = \rho_{ij}\sigma_i\sigma_j, \qquad i = 1, \ldots, m; \qquad j = 1, \ldots, m.$$

Denote the determinant of Σ by $|\Sigma|$ and the inverse by Σ^{-1}. Then the probability density function of $\mathbf{x}$ is given by

$$f(\mathbf{x}) = (2\pi)^{-m/z}|\Sigma|^{-1/2}\exp\{-\tfrac{1}{2}(\mathbf{x}-\boldsymbol{\theta})'\Sigma^{-1}(\mathbf{x}-\boldsymbol{\theta})\}. \qquad (3.6)$$

Consider the special case of $p = 2$. The multivariate Normal distribution is then called the bivariate Normal distribution. The vector parameters may be written out in full.

$$\boldsymbol{\theta} = \begin{pmatrix} \theta_1 \\ \theta_2 \end{pmatrix}$$

$$\Sigma = \begin{pmatrix} \sigma_1^2 & \rho\sigma_1\sigma_2 \\ \rho\sigma_1\sigma_2 & \sigma_2^2 \end{pmatrix} \qquad (3.7)$$

$$\Sigma^{-1} = \frac{1}{1-\rho^2}\begin{pmatrix} \sigma_1^{-2} & -\rho/\sigma_1\sigma_2 \\ -\rho/\sigma_1\sigma_2 & \sigma_2^{-2} \end{pmatrix}$$

$$|\Sigma|^{1/2} = \sigma_1\sigma_2\sqrt{(1-\rho^2)}$$

The matrix Σ^{-1} is sometimes denoted Ω (see Section 4.5.5). Notice that

$$(\mathbf{x}-\boldsymbol{\theta})'\Sigma^{-1}(\mathbf{x}-\boldsymbol{\theta}) = \left\{\frac{(x_1-\theta_1)^2}{\sigma_1^2} - 2\rho\frac{(x_1-\theta_1)(x_2-\theta_2)}{\sigma_1^2\sigma_2^2} + \frac{(x_2-\theta_2)^2}{\sigma_2^2}\right\}\bigg/(1-\rho^2).$$

The bivariate Normal density function may then be written as

$$f(x_1,x_2) = \frac{1}{2\pi\sigma_1\sigma_2\sqrt{(1-\rho^2)}}$$

$$\times\exp\left[-\frac{1}{2(1-\rho^2)}\left\{\frac{(x_1-\theta_1)^2}{\sigma_1^2} - 2\rho\frac{(x_1-\theta_1)(x_2-\theta_2)}{\sigma_1^2\sigma_2^2} + \frac{(x_2-\theta_2)^2}{\sigma_2^2}\right\}\right].$$

For the special case in which $\theta_1 = \theta_2 = 0$,

$$f(x_1,x_2) = \frac{1}{2\pi\sigma_1\sigma_2\sqrt{(1-\rho^2)}}\exp\left\{-\frac{1}{2(1-\rho^2)}\left(\frac{x_1^2}{\sigma_1^2} - 2\rho\frac{x_1x_2}{\sigma_1^2\sigma_2^2} + \frac{x_2^2}{\sigma_2^2}\right)\right\}.$$

Applications are given in Sections 4.4.5 and 4.5.3.

4

Historical Review

4.1 EARLY HISTORY

The earliest use of probabilistic reasoning in legal decision making, albeit in a somewhat rudimentary form, appears to have been over 18 centuries ago by Jewish scholars in Babylon and Israel writing in the *Talmud* (Zabell, 1976, in a review of Rabinovitch, 1973). For example, if nine stores in a town sold kosher meat and one sold non-kosher meat, then a piece of meat which is found, at random, in the town is presumed to be kosher and thus ritually permissible to eat since it is assumed to have come from one of the shops in the majority (Rabinovitch, 1969). However, consider the following quote from the *Talmud*:

> 'All that is stationary (fixed) is considered half and half. . . . If nine shops sell ritually slaughtered meat and one sells meat that is not ritually slaughtered and he bought in one of them and does not know which one, it is prohibited because of the doubt; but if meat was found in the street, one goes after the majority.'
> (Kethuboth 15a, quoted in Rabinovitch, 1969.)

The reasoning seems to be as follows. If the question arises at the source of the meat (i.e., in the shops) the odds in favour of kosher meat are not really 9 to 1. The other nine shops are not considered—the piece of meat certainly did not come from any of them. There are, hence, only two possibilities: the meat is either kosher or it is not. The odds in favour of it being kosher is evens. However, if meat is found outside the shops (e.g., in the street) the probability that it came from any one of the ten shops is equal for each of the ten shops. Thus, the probability that it is kosher is 0.9.

The works of Cicero (*De Inventione* and *Rhetorica ad Herennium*) and Quintillian (*Institutio Oratoria*) among others are cited by Garber and Zabell (1979). Garber and Zabell also quote an example from James Bernoulli's *Ars Conjectandi* (1713, Part 4, Chapter 2) which is of interest given the examples in Section 1.6.7 of updating probabilities.

The discussion by Bernoulli is as follows. One person, Titius, is found dead on the road. Another, Maevius, is accused of committing the murder. There are various pieces of evidence in support of this accusation.

1. It is well known that Maevius regarded Titius with hatred. (This is evidence of motive: hatred could have driven Maevius to kill.)
2. On interrogation, Maevius turned pale and answered apprehensively. (This is evidence of effect: the paleness and apprehension could have come from his own knowledge of having committed a crime.)
3. A blood stained sword was found in Maevius' house (This is evidence of a weapon).
4. On the day Titius was slain, Maevius travelled over the road (This is evidence of opportunity.)
5. A witness, Gaius, alleges that on the day before the murder he had interceded in a dispute between Titius and Maevius.

Later (Chapter 3 of Part 4 of *Ars Conjectandi*) James Bernoulli (1713) discussed how to calculate *numerically* the weight which should be afforded a piece of evidence or proof.

> 'The degree of certainty or the probability which this proof generates can be computed from these cases by the method discussed in the first part (i.e., the ratio of favourable to total cases) just as the fate of the gamblers in games of chance are accustomed to be investigated.' (Garber and Zabell, 1979, p. 44.)

Garber and Zabell (1979) then go on to say

> 'What is new in the "*Ars Conjectandi*" is not its notation of evidence—which is based on the rhetorical treatment of circumstantial evidence—but its attempt to *quantify* such evidence by means of the newly developed calculus of chances.' (p. 44.)

Thus it is that over two hundred years ago, consideration was being given to methods of evaluating evidence numerically.

A long discussion of *Ars Conjectandi*, Part 4, is given by Shafer (1978). The distinction is drawn between pure and mixed arguments. A *pure argument* is one which proves a thing in certain cases in such a way as to prove nothing positively in other cases. A *mixed argument*, on the other hand, is one which proves a thing in some cases in such a way that they prove the contrary in the remaining cases. Shafer discusses an example of this from Part 4 of *Ars Conjectandi.*

A man is stabbed with a sword in the midst of a rowdy mob. It is established by the testimony of trustworthy men who were standing at a distance that the crime was committed by a man in a black cloak. It is found that one person, by the name of Gracchus, and three others in the crowd were wearing cloaks of that colour. This is an argument that the murder was committed by Gracchus but it is a *mixed* argument. In one case it proves his guilt, in three cases his innocence, according to whether the murder was perpetrated by himself or one of the other three. If one of those three perpetrated the murder then Gracchus is supposed innocent.

However, if at a subsequent hearing, Gracchus went pale, this is a *pure* argument. If the change in his pallor arose from a guilty conscience it is indicative of his guilt. If it arose otherwise it does not prove his innocence; it could be that Gracchus went pale for a different reason but that he is still the murderer.

Shafer (1978) draws an analogy between these two kinds of argument and his mathematical theory of evidence (Shafer, 1976) and belief functions (see Section 1.2). In that theory a probability p is assigned to a proposition and a probability q to its negation, or complement, such that $0 \leqslant p \leqslant 1$, $0 \leqslant q \leqslant 1$, $p + q \leqslant 1$. It is not necessarily the case that $p + q = 1$, in contradiction to (2.1). There are then three possibilities:

- $p > 0$, $q = 0$ implies the presence of evidence in favour of the proposition and the absence of evidence against it;
- $p > 0$, $q > 0$ implies the presence of evidence on both sides, for and against, the proposition;
- $p > 0$, $q > 0$, $p + q = 1$ (additivity) occurs only when there is very strong evidence both for and against the proposition.

Only probabilities which satisfy the additivity rule (2.1) are considered in this book.
Socrates also discussed the use of probabilistic ideas in law.

> 'Men care nothing about truth but only about conviction and this is based on probability'

is a quote given by Sheynin (1974). Sheynin also quotes Aristotle (*Rhetorica*, 1376a, 19) as saying

> 'If you have no witnesses ... you will argue that the judges must decide from what is probable ... If you have witnesses, and the other man has not, you will argue that probabilities cannot be put on their trial and that we could do without the evidence of witnesses altogether if we need do no more than balance the pleas advanced on either side.'

Sheynin (1974) also mentions that probability in law was discussed by Thomas Aquinas (*Treatise on Law*, Question 105, Article 2, *Great Books*, volume 20, p. 314) who provides a comment on collaborative evidence.

> 'In the business affairs of men, there is no such thing as demonstrative and infallible proof and we must contend with a certain conjectural probability. .... Consequently, although it is quite possible for two or three witnesses to agree to a falsehood, yet it is neither easy nor probable that they succeed in so doing; therefore their testimony is taken as being true.'

James Bernoulli (1713) gave a probabilistic analysis of the cumulative force of circumstantial evidence. His nephew, Nicholas Bernoulli (1709), applied the calculus of probabilities to problems including the presumption of death, the value of annuities, marine insurance, the veracity of testimony and the probability of innocence; see Fienberg (1989).

   The application of probability to the verdicts by juries in civil and criminal trials was discussed by Poisson (1837) and there is also associated work by Condorcet (1785), Cournot (1838) and LaPlace (1886). The models developed by Poisson have been put in a modern setting by Gelfand and Solomon (1973).

Two early examples of the use of statistics to query the authenticity of signatures on wills were given by Mode (1963). One of these, the Howland will case from the 1860's has been discussed also by Meier and Zabell (1980). This case is probably the earliest instance in American law of the use of probabilistic and statistical evidence. The evidence was given by Professor Benjamin Peirce, Professor of Mathematics at Harvard University, and by his son Charles, then a member of staff of the United States Coast Survey. The evidence related to the agreement of thirty downstrokes in a contested signature with those of a genuine signature. It was argued that the probability of this agreement if the contested signature were genuine was extremely small. Hence the contested signature was a forgery.

More recent attempts to evaluate evidence are reviewed here in greater detail.

4.2   DREYFUS

This example concerns the trial of Dreyfus in France at the end of the last century. Dreyfus, an officer in the French Army assigned to the War Ministry, was accused in 1894 of selling military secrets to the German military attaché. Part of the evidence against Dreyfus centred on documents, admitted to have been written by him, and said by his enemies to contain cipher messages. This assertion was made because the letters of the alphabet did not occur in these documents in the proportions in which they were known to occur in average French prose. The proportions observed had a very small probability of occurring (see Tribe, 1971). Though it was pointed out to the lawyers that the most probable proportion of letters was itself highly improbable, this point was not properly understood.

A simple example from coin tossing will suffice to explain what is meant by the phrase 'the most probable proportion of letters was itself highly improbable'. Consider a fair coin, that is one in which the probabilities of a head and of a tail are equal at $1/2$. If the coin is tossed ten thousand times, the expected number of heads is 5000 (see Section 3.3.3 with $n = 10\,000$, $p = 1/2$) and this is also the most probable outcome. However, the probability of 5000 heads, as distinct from 4999 or 5001 or any other number, is $\simeq 0.008$ or 1 in 125 which is a very low probability. The most probable outcome is, itself, improbable. The situation, of course, would be considerably enhanced given all the possible choices of combinations of letters in French prose in Dreyfus' time.

It is not difficult to see where the lawyers' logic had failed them. If Dreyfus were innocent $(\bar{G})$ the combination of letters (E) which he had used would be extremely unlikely—$Pr(E|\bar{G})$ would be very small. Therefore they concluded that Dreyfus must have deliberately chosen the letters he did as a cipher and so must be a spy—$Pr(\bar{G}|E)$ must be very small. The lawyers did not see that any other combination of letters would also be extremely unlikely and that the particular combination used by Dreyfus was of no great significance. This is an early example of the fallacy of the transposed conditional (Section 2.3.1).

4.3 PEOPLE v. COLLINS

The Dreyfus case is a rather straightforward abuse of probabilistic ideas though the fallacy of which it is an example still occurs. It is easy now to expose the fallacy through consideration of the odds form of Bayes' Theorem (see Section 2.5.4). At the time, however, the difficulty of having the correct reasoning accepted had serious and unfortunate consequences for Dreyfus. A further example of the fallacy occurred in a case which has achieved a certain notoriety in the probabilistic legal literature, namely that of *People v. Collins* (California Reporter, 1968, **66**, 497; Fairley and Mosteller, 1977). In this case, probability values, for which there was no objective justification, were quoted in court.

Briefly, the crime was as follows. An old lady was pushed to the ground in an alley-way in Los Angeles by someone whom she neither saw nor heard. She managed to look up and saw a young man running from the scene. Immediately after the incident the lady discovered her purse was missing. A man at the end of the alley noticed a woman, whom he later described as a Caucasian with her hair in a blonde ponytail, run out of the alley and enter a yellow automobile parked across the street. The car was driven by a male Negro wearing a moustache and a beard.

A couple answering this description were eventually arrested and brought to trial. The prosecutor called as a witness an instructor of mathematics at a state college in an attempt to bolster the identifications. This witness testified to the product rule for multiplying together the probabilities of independent events (the third law of probability (1.3), Section 1.6.4). The third law may be extended to a set of n independent events and also to take account of conditional probabilities. Thus, if $E_1, E_2, \ldots, E_n$ are mutually independent pieces of evidence and $\bar{G}$ denotes the hypothesis of innocence then

$$Pr(E_1 E_2 \ldots E_n | \bar{G}) = Pr(E_1 | \bar{G}) Pr(E_2 | \bar{G}) \ldots Pr(E_n | \bar{G}). \tag{4.1}$$

In words, this states that, for mutually independent events, the probability that they all happen is the product of the probabilities of each individual event happening.

The instructor of mathematics then applied this rule to the characteristics as testified to by the other witnesses. Values to be used for the probabilities of the individual characteristics were suggested by the prosecutor without any justification, a procedure which would not now go unchallenged. The jurors were invited to choose their own values but, naturally, there is no record of whether they did or not. The individual probabilities suggested by the prosecutor are given in Table 4.1. Using the product rule for independent characteristics, the prosecutor calculated the probability that a couple selected at random from a population would exhibit all these characteristics as 1 in 12 million ($10 \times 4 \times 10 \times 3 \times 10 \times 1000 = 12\,000\,000$).

Table 4.1   Probabilities for various characteristics of the couple observed in the case of *People v. Collins*

Evidence	Characteristic	Probability
E_1	Partly yellow automobile	1/10
E_2	Man with moustache	1/4
E_3	Girl with ponytail	1/10
E_4	Girl with blonde hair	1/3
E_5	Negro man with beard	1/10
E_6	Inter-racial couple in car	1/1000

The accused were found guilty. This verdict was overturned on appeal for two reasons.

(a) The statistical testimony lacked an adequate foundation both in evidence and in statistical theory.
(b) The testimony and the manner in which the prosecution used it distracted the jury from its proper function of weighing the evidence on the issue of guilt.

The first reason refers to the lack of justification offered for the choice of probability values and the assumption that the various characteristics were independent. As an example of this latter point, an assumption of independence assumes that the propensity of a man to have a moustache does not affect his propensity to have a beard.

The second reason still has considerable force today. When statistical evidence is presented, great care has to be taken that the jury is not distracted from 'its proper function of weighing the evidence on the issue of guilt'.

The fallacy of the transposed conditional is again evident here. The evidence is $(E_1, E_2, \ldots, E_6)$ and $Pr(E_1 E_2 \ldots E_6 | \bar{G})$ is extremely small (1 in 12 million). The temptation for a juror to interpret this figure as a probability of innocence is very great.

4.4   DISCRIMINATING POWER

4.4.1   Derivation

How good is a method at distinguishing between two samples of material from different sources? If a method fails to distinguish between two samples, how strong is this as evidence that the samples come from the same source? Questions like these and the answers to them were of considerable interest to forensic scientists in the late 1960's and in the 1970's; see, for example, theoretical work in Parker (1966, 1967), Jones (1972) and Smalldon and Moffat (1973). Experimental attempts to answer these questions are described in Tippett *et al.* (1968)

for fragments of paint, in Gaudette and Keeping (1974) for human head hairs and for shoeprints in Groom and Lawton (1987).

Two individuals are selected at random from some population. The probability that they are found to match with respect to some characteristic (e.g., blood phenotype, paint fragments on clothing, head hairs) is known as the *probability of non-discrimination* or the *probability of a match, PM*. The complementary probability, the probability they are found not to match with respect to this characteristic is known as the *probability of discrimination* (Jones, 1972) or *discriminating power, DP* (Smalldon and Moffat, 1973). The idea was first applied to problems concerning ecological diversity (Simpson, 1949) and later to blood group genetics (Fisher, 1951). See also Jeffreys *et al.* (1987) for an application to DNA profiles.

Consider a population with a blood group system S in which there are k phenotypes and in which the jth phenotype has relative frequency p_j, such that $p_1 + \cdots + p_k = 1$ from the second law of probability for mutually exclusive and exhaustive events, Section 1.6.6. Two people are selected at random from this population such that their phenotypes may be assumed independent. What is the probability of a match of phenotypes between the two people in the S system?

Let the two people be called C and D. Let C_1 and D_1 be the events that C and D are of phenotype 1, respectively. Then

$$Pr(C_1) = Pr(D_1) = p_1.$$

Then the probability of C_1 and of D_1 is given by

$$Pr(C_1 D_1) = Pr(C_1) \times Pr(D_1) = p_1^2$$

by the third law of probability (1.3) applied to independent events. Thus, the probability they are both of phenotype 1 is p_1^2. In general, let C_j, D_j be the events that C, D are of phenotype j ($j = 1, \ldots, k$). The probability that the individuals selected at random match on phenotype j, is given by

$$Pr(C_j D_j) = Pr(C_j) \times Pr(D_j) = p_j^2.$$

The probability of a match on *any* phenotype is the disjunction of k mutually exclusive events, the matches on phenotypes $1, \ldots, k$, respectively. Let Q be the probability PM of a match. Then

$$Q = Pr(C_1 D_1 \text{ or } C_2 D_2 \text{ or } \ldots \text{ or } C_k D_k)$$
$$= Pr(C_1 D_1) + Pr(C_2 D_2) + \cdots + Pr(C_k D_k)$$
$$= p_1^2 + p_2^2 + \cdots + p_k^2, \tag{4.2}$$

by the second and third laws of probability (1.2 and 1.3); see also (1.1) of Example 1.1. The discriminating power, or probability of discrimination, is $1 - Q$.

Example 4.1   Selvin *et al.* (1983) give an example using the *ABO* system with four phenotypes ($k = 4$) and gene frequencies from Grunbaum *et al.* (1980), given in Table 4.2:

$$Q = 0.441^2 + 0.417^2 + 0.104^2 + 0.039^2 = 0.381.$$

The discriminating power $DP = 1 - Q = 0.619$.

Notice the following point with reference to the comparison of blood samples. The blood groups of two blood samples from unknown sources are to be compared. What is the most likely outcome for the two blood groups? The most common blood group is O. The probability the samples are both O is $0.441^2 = 0.194$ using the figures in Table 4.2. However, this is not the most common outcome for the blood groups of two samples. Identify the two sources as C and D. The probability that C is of group O and that D is of group A is 0.441×0.417. Similarly, the probability that C is of group A and that D is of group O is 0.417×0.441. The probability that the blood groups of the two samples are O and A is $2 \times 0.441 \times 0.417 = 0.368$. The most common outcome when grouping two blood samples from unknown sources is OA, not OO.

4.4.2   Evaluation of evidence by discriminating power

The approach using discriminating power has implications for the assessment of the value of forensic evidence. If two samples of material (e.g., two blood stains, two sets of paint fragments, two groups of human head hairs) are found to be indistinguishable, it is of interest to know if this is forensically significant. If a system has a high value of Q it implies that a match between samples of materials from two different sources under this system is quite likely. It is intuitively reasonable that a match under such a system will not be very significant. Conversely, if a system has a very low value of Q, a match will be forensically significant. Data from the Strathclyde region of Scotland for which there is a discriminating power of 0.602 are given by Gettinby (1984). He interprets this to mean that 'in 100 cases where two blood samples come from different people then, on average, 60 will be identifiable as such'.

Limits for *PM* may be determined (Jones, 1972). First, note that $p_1 + \cdots + p_k = 1$ and that $0 \leqslant p_j \leqslant 1$ ($j = 1, \ldots, k$) from the first law of probability (1.4). Thus $p_j^2 \leqslant p_j$ ($j = 1, \ldots, k$) and so $Q = p_1^2 + \cdots + p_k^2 \leqslant 1$; i.e., Q can never be greater than 1 (and equal to 1 if and only if one of the p_j's is 1, and hence the rest are 0). A value

Table 4.2   Phenotypic frequencies in the *ABO* system, from Grunbaum *et al* (1980)

Phenotype (j)	O	A	B	AB
Frequency (p_j)	0.441	0.417	0.104	0.039

of Q equal to 1 implies that all members of the population fall into the same category; the discriminating power is zero.

Now, consider the lower bound. It is certainly no less than zero. Suppose the characteristic of interest (h_0 say) divides the system into k classes of equal probability $1/k$ so that $p_j = p_{0j} = 1/k$ ($j = 1, \ldots, k$). Then

$$Q = Q_0 = p_{01}^2 + p_{02}^2 + \cdots + p_{0k}^2$$

$$= \frac{1}{k^2} + \frac{1}{k^2} + \cdots + \frac{1}{k^2}$$

$$= \frac{1}{k}.$$

Consider another characteristic (h_1 say) which divides the system into k classes of unequal probability such that

$$p_j = p_{1j} = \frac{1}{k} + \varepsilon_j; \qquad j = 1, \ldots, k.$$

Since $\Sigma_{j=1}^k p_j = 1$ it can be inferred that $\Sigma_{j=1}^k \varepsilon_j = 0$. Thus, for h_1,

$$Q = Q_1 = p_{11}^2 + p_{12}^2 + \cdots + p_{1k}^2$$

$$= \Sigma_{j=1}^k \left(\frac{1}{k} + \varepsilon_j \right)^2$$

$$= \Sigma_{j=1}^k \left(\frac{1}{k^2} + \frac{2\varepsilon_j}{k} + \varepsilon_j^2 \right)$$

$$= \frac{1}{k} + \frac{2}{k} \Sigma_{j=1}^k \varepsilon_j + \Sigma_{j=1}^k \varepsilon_j^2$$

$$= \frac{1}{k} + \Sigma_{j=1}^k \varepsilon_j^2 \text{ since } \Sigma_{j=1}^k \varepsilon = 0$$

$$\geqslant Q_0$$

since $\Sigma_{j=1}^k \varepsilon_j^2$ is never negative {and equals zero if and only if $\varepsilon_1 = \varepsilon_2 = \cdots = \varepsilon_k = 0$; i.e., if $p_{0j} = p_{1j}$ ($j = 1, \ldots, k$)}. Thus PM takes values between $1/k$ and 1 where k is the number of categories in the system. The probability of a match is minimised, and the discriminating power is maximised when the class probabilities are all equal. This is confirmation of a result which is intuitively reasonable, namely that if a choice has to be made among several techniques as to which to implement, then techniques with greater variability should be preferred over those with lesser variability.

As an example of the application of this result, note that in the *ABO* system above that $k = 4$ and $1/k = 0.25$. Thus, *PM* cannot be lower than 0.25 for this *ABO* system and the discriminating power cannot be greater than 0.75. Notice also that the minimum value $(1/k)$ of *PM* decreases as k increases. Discriminating power increases as k increases, a result which is intuitively attractive.

The above calculations assume that the total population size (N say) of interest is very large and that for at least one p_j, p_j^2 is much greater than $1/N$. Failure of these assumptions can lead to misleading results as described by Jones (1972) with reference to the results of Tippett *et al* (1968) on paint fragments; see Examples 4.2 and 4.3.

The experiment described by Tippett *et al* (1968) compared two thousand samples of paint fragments, pairwise. For various reasons, the number of samples was reduced to 1969, all from different sources. These were examined by various tests and only two pairs of samples from different sources were found to be indistinguishable. The total number of pairs which can be picked out at random is $\frac{1}{2} \times 1969 \times 1968 = 1937\,496$. Two pairs of samples were found to agree with each other. The probability of picking a pair of fragments at random which are found to be indistinguishable is thus determined empirically as $2/1937\,496 = 1/968\,748$.

This probability is an estimate of the probability of a match (*PM*). The method by which it was determined is extremely useful in situations such as the one described by Tippett *et al.* (1968) in which frequency probabilities are unavailable and, indeed, for which a classification system has not been devised. The extremely low value $(1/968\,748)$ of the *PM* demonstrates the extremely high evidential value of the methods used by the authors. The conclusion from this experiment is that these methods are very good at differentiating between paints from different sources. Low values of *PM* were also obtained in work on head hairs (Gaudette and Keeping, 1974) and footwear (Groom and Lawton, 1987).

The equivalence of the theoretical and empirical approaches to the determination of *PM* can be verified numerically using the *ABO* system with phenotypic frequencies as given in Table 4.2. Assume there is a sample of 1000 people with blood group frequencies in the proportions in Table 4.2. All possible pairs of people in this sample are considered and their phenotypes compared. There are $\frac{1}{2}(1000 \times 999) = 499\,500\,(= P$, say) different pairings. Of these pairings there are the following numbers of matches for:

- phenotype *O*: $441 \times 440/2$,
- phenotype *A*: $417 \times 416/2$,
- phenotype *B*: $104 \times 103/2$,
- phenotype *AB*: $39 \times 38/2$.

There are, thus, $M = \{(441 \times 440) + (417 \times 416) + (104 \times 103) + (39 \times 38)\}/2 = 189\,853$ pairings of people who have the same phenotype. The probability of a match, by this numerical method, is then $M/P = 189\,853/499\,500 = 0.3801$.

The probability of a match, as given by PM, is $PM = p_1^2 + \cdots + p_4^2 = 0.441^2 + 0.417^2 + 0.104^2 + 0.039^2 = 0.3807$. The approximate equality of these two values is not a coincidence as a study of the construction of the ratio M/P shows.

The probability PM is sometimes called an *average probability* (Aitken and Robertson, 1987). An average probability provides a measure of the effectiveness of a particular type of transfer evidence at distinguishing between two randomly selected individuals (Thompson and Williams, 1991). In the context of blood stains, it is so-called because it is the average of the probabilities that an innocent person will be found to have a phenotype which matches that of a crime stain. These probabilities are of the type considered in Example 1.1. For example, if the crime stain were of type O, the probability that an innocent suspect matches this is just the probability that he is of type O, namely 0.441. A similar argument gives probabilities of 0.427, 0.104 and 0.039 for matches between crime stain phenotypes and that of an innocent person of types A, B and AB, respectively. The average probability is the average of these four probabilities, weighted by their relative frequencies in the population, which are 0.441, 0.427, 0.104 and 0.039, respectively. The average probability is then just PM, given by $(0.441 \times 0.441) + (0.427 \times 0.427) + (0.104 \times 0.104) + (0.039 \times 0.039) = 0.441^2 + 0.427^2 + 0.104^2 + 0.039^2 = 0.3807$.

4.4.3 Finite samples

The relationship between the general result for a population, which is conceptually infinite in size, and a sample of finite size is explained by Jones (1972). Consider a test to distinguish between k classes $C_1, \ldots, C_k$. A sample of n individuals is taken from the relevant population. The number of individuals in each class are $c_1, c_2, \ldots, c_k$ with $\Sigma_{j=1}^{k} c_j = n$. An estimate of the probability that a randomly selected individual will be in class C_j is $\hat{p}_j = c_j/n, j = 1, 2, \ldots, k$.

There are $n(n-1)/2$ possible pairings of individuals. For any particular class j, the number of pairings of individuals within the class is $c_j(c_j - 1)/2, j = 1, 2, \ldots, k$. Thus, the overall proportion of pairings which result in a match is

$$\hat{Q} = \{\Sigma_{j=1}^{k} c_j(c_j - 1)\}/\{n(n-1)\}.$$

Then

$$\hat{Q} = (\Sigma_{j=1}^{k} c_j^2 - \Sigma_{j=1}^{k} c_j)/\{n(n-1)\}$$

$$= (\Sigma_{j=1}^{k} c_j^2 - n)/(n^2 - n) \quad \text{since} \quad \Sigma_{j=1}^{k} c_j = n,$$

$$= \{\Sigma_{j=1}^{k} (c_j^2/n^2) - 1/n\}/(1 - 1/n)$$

$$= (\Sigma_{j=1}^{k} \hat{p}_j^2 - 1/n)/(1 - 1/n). \tag{4.3}$$

When the class frequencies are known, the probability of a match PM is given by Q (4.2). The result above (4.3) gives an exact expression for the probability of a match for any given sample for all values of n and $\{\hat{p}_j, j = 1, \ldots, k\}$. As n increases towards the population size it is to be expected that the observed sample probability will converge to the population probability. For n large it can be seen that $\hat{Q}$ tends to

$$\hat{Q}_1 = \sum_{j=1}^{k} \hat{p}_j^2,$$

since $1/n$ becomes vanishingly small. As $\hat{p}_j$ tends to p_j so $\hat{Q}_1$ will tend to Q. However, as well as n being large, it is necessary for at least one of the $\hat{p}_j^2$ to be much greater than $1/n$ in order that $\Sigma \hat{p}_j^2 - 1/n \simeq \Sigma p_j^2$.

Two examples (Jones, 1972) are given in which the probability of a match estimated by $\hat{Q}_1 = \Sigma \hat{p}_j^2$ is not a good approximation to the probability of a match estimated by $\hat{Q}$. The true probability of a match, $Q = \Sigma p_j^2$ will not often be known exactly, except in situations like blood grouping systems where the sample sizes are extremely large and the $\{p_j\}$ are known accurately.

Example 4.2   Small sample size n   If n is small, then $1/n$ is not very small compared to 1. Consider four playing cards, of which two are red (R_1, R_2: Category 1) and two are black (B_1, B_2: Category 2). Thus, $n = 4$, $c_1 = c_2 = 2$, $p_1 = \hat{p}_1 = p_2 = \hat{p}_2 = 1/2$ and $\hat{Q}_1 = 1/4 + 1/4 = 1/2$. Note that $1/n = 1/4$ which is not very much less than 1. There are six possible pairings of cards ($R_1 R_2$, $R_1 B_1$, $R_1 B_2$, $R_2 B_1$, $R_2 B_2$, $B_1 B_2$) of which two result in a match ($R_1 R_2$, $B_1 B_2$). Thus $\hat{Q} = 1/3$ which may be verified from $\hat{Q} = (\Sigma \hat{p}_j^2 - 1/n)/(1 - 1/n) = (1/2 - 1/4)/(1 - 1/4) = 1/3$. The failure of $1/n$ to be very small has led to a discrepancy between $\hat{Q}$ and $\hat{Q}_1$.

Example 4.3   Very small values of p_j^2   Consider the paper by Tippett *et al.* (1968). There, 1969 sets of paint fragments were considered ($n = 1969$). Two pairs were found to be indistinguishable. Label these pairs as belonging to classes 1 and 2. The other paint fragments may be said to belong to 1965 different classes labelled $3, \ldots, 1967$, each with only one member so that $\hat{p}_1 = 2/1969$, $\hat{p}_2 = 2/1969$, $\hat{p}_3 = \cdots = \hat{p}_{1967} = 1/1969$. Then

$$\hat{Q}_1 = \left(\frac{2}{1969}\right)^2 + \left(\frac{2}{1969}\right)^2 + \left(\frac{1}{1969}\right)^2 + \cdots + \left(\frac{1}{1969}\right)^2$$

$$= \frac{1973}{1969^2}$$

$$\simeq \frac{1}{1965},$$

whereas

$$\hat{Q} = \left(\frac{1973}{1969^2} - \frac{1}{1969} \right) \bigg/ \left(1 - \frac{1}{1969} \right)$$

$$= \frac{4}{1969 \times 1968}$$

$$= \frac{1}{968\,748},$$

agreeing with the earlier result obtained by the authors. Here the approximate result $\hat{Q}_1$ is very inaccurate because no $\hat{p}_j^2$ is very much greater than $1/n$. In fact, the largest $\hat{p}_j^2$ is $(2/1969)^2$ which is smaller than $1/n$.

4.4.4 Combination of independent systems

Consider Q from equation (4.2). This is the probability of finding a match between two individuals selected at random using a particular classification system. Suppose now that there are m independent systems with corresponding PM values $Q_1, \ldots, Q_m$. The probability of finding a pair which match on all m tests is $PM_m = \prod_{l=1}^{m} Q_l$. The probability of being able to distinguish between two individuals using these m tests is therefore

$$DP_m = 1 - \prod_{l=1}^{m} Q_l.$$

In a comparison of the phenotype frequencies between Africans and East Indians, Thompson and Williams (1991) gave frequency results for the *ABO, EAP, GLO* and *PGM* systems in Trinidad and Tobago. The calculations for the discriminat-

Table 4.3 Phenotypic frequencies in *ABO* system for Africans and East Indians and the probability Q_{ABO} of a match (from Thompson and Williams, 1991)

Phenotype	Frequency	
	African	East Indian
A	0.228	0.206
B	0.198	0.342
O	0.539	0.382
AB	0.035	0.069
Q_{ABO}	0.383	0.310

Table 4.4 Phenotypic frequencies in *GLO* system for Africans and East Indians and the bability Q_{GLO} of a match (from Thompson and Williams, 1991)

Phenotype	Frequency African	East Indian
1	0.075	0.091
2–1	0.398	0.421
2	0.527	0.487
Q_{GLO}	0.442	0.423

ing power for the combination of the *ABO* and *GLO* systems are given below. Further details are given in Thompson and Williams (1991).

The phenotypic frequencies for the *ABO* and *GLO* systems for Africans and East Indians are given in Tables 4.3 and 4.4. The probability that two blood samples match on both criteria is

$$PM_2 = Q_{ABO} \times Q_{GLO}$$

$$= 0.169 \text{ (African)},$$

$$= 0.131 \text{ (East Indian)}.$$

The discriminating power is

$$DP_2 = 0.831 \text{ (African)},$$

$$= 0.869 \text{ (East Indian)}.$$

Results for Great Britain are given by Jones (1972) using figures given by Mourant *et al.* (1958) and Stedman (1972).

There is further discussion in Section 5.5 concerning the determination of matching probabilities when there may be errors in the classification.

4.4.5 Correlated attributes

Discrete attributes

Consider the hair example of Gaudette and Keeping (1974) in more detail. (A similar argument holds for the paint example of Tippett *et al.*, 1968.) A series of 366 630 pairwise comparisons between hairs from different individuals were made. Nine pairs of hairs were found to be indistinguishable. These results were used to provide an estimate of the probability that a hair taken at random from one individual, A say, would be indistinguishable from a hair taken at random from another individual, B say, namely $9/366\,630$ or $1/40\,737$. It was then argued that if nine dissimilar hairs were independently chosen to represent the hairs on the scalp of individual B, the chance that a single hair for A is distinguishable from all nine of B's may be taken as $[1 - (1/40\,737)]^9$ which is approximately $1 - (1/4\,500)$. The complementary probability, the probability that a single hair from A is indistinguishable from at least one of B's hairs is $1/4500$. This probability provides, in some sense, a measure of the effectiveness of human hair comparison in forensic hair investigations.

There are various criticisms, though, which can be made regarding this approach, some details of which are in Aitken and Robertson (1987). Comments on the Gaudette–Keeping study are also made in Fienberg (1989).

First note that the assumption of independence of the nine hairs used in the calculation is not an important one. The use of an inequality known as the *Bonferroni inequality* gives an upper bound for the probability investigated by the authors of $1/4526$ (Gaudette, 1982). The Bonferroni inequality states that the probability that at least one of several events occurs is never greater than the sum of the probabilities of the occurrences of the individual events. Denote the events by $R_1, R_2, \ldots, R_n$, then the inequality states that

$$Pr(\text{at least one of } R_1, \ldots, R_n \text{ occurs}) \leqslant Pr(R_1) + \cdots + Pr(R_n).$$

Gaudette and Keeping (1974) compared each of nine hairs, known to be from one source with one hair known to be from another source. The events $R_1, \ldots, R_9$, where $n = 9$, correspond to the inability to distinguish each of the nine hairs from the single hair. The probability of interest is the probability of at least one indistinguishable pair in these nine comparisons. This is the probability that at least one of $R_1, \ldots, R_9$ occurs. Using the Bonferroni inequality, it can be seen that this probability is never greater than the sum of the individual probabilities. From above, these individual probabilities are all equal to $1/40\,737$. The sum of the nine of them then equals $9/40\,737$, which is $1/4526$. This is very close to the figure of $1/4500$ quoted from the original experiment. Even if independence is not assumed, there is very little change in the probability figure quoted as a measure of the value of the evidence.

An important criticism, though, is the following. One probability of interest to the court is the probability that a hair found at a crime scene belonged to a suspect. Other probabilities of interest, the relevance of which will be explained in more detail later in Chapter 5, are the probability of the evidence of the similarity of the hairs (crime and suspect) if they came from the same origin and the probability of the evidence of the similarity of the hairs if they came from different origins. Gaudette and Keeping (1974) provided an estimate of the probability that hairs 'selected at random' from two individuals are indistinguishable. This probability is an average probability (Section 4.4.2). It can be used as a broad guideline to indicate the effectiveness of hair identification in general.

However, the use of the figure 1/4500 as the value of the evidence in a particular case could be very misleading.

The average probability is the probability that two individuals chosen at random will be indistinguishable with respect to the trait under examination. However, in a particular investigation one sample is of known origin (in this case, the sample of hair from the suspect, which is the source sample), the other (the crime or receptor sample) is not. If the suspect is not the criminal, someone else is and the correct probability of interest is the probability that a person chosen at random (which is how the criminal must be considered) from some population will have hair which is similar to that of the suspect—see Chapter 6 for a more detailed discussion.

Fienberg (1989) also makes the point that, 'even if we interpret 1/4500 . . . as the probability of a match given a comparison of a suspected hair with a sample from a different individual we would still need the probability of a match given a sample from the same individual as well as *a priori* probabilities of "same" and "different".'

Continuous measurements

The problem of interpreting correlated continuous attributes in which the underlying distribution is multivariate Normal (Section 3.4.1) was discussed by Smalldon and Moffat (1973). First, consider a set of p uncorrelated continuous attributes with measurements $\{x_l, l = 1, \ldots, p.\}$ and corresponding density functions $f(x_1), \ldots, f(x_p)$. An estimate of the probability that two individuals chosen at random from the population will match, in some way which has to be defined, on all p attributes is required.

For attribute l, the probability that the measurement on the first individual will lie in a small interval t of width dx_l about measurement x_l is $f(x_l)dx_l$. This may be understood intuitively by realising that $f(x_l)$ is the height of the probability density curve at the point x_l. For example, see Figure 3.1; the height of the Normal density curve of mean zero and variance 1250 at the point $x_l = 30$ is 0.0079 from (3.5). The probability $f(x_l)dx_l$ is the area of a narrow rectangle of height $f(x_l)$ and width dx_l. This is used as an approximation to the probability that the measurement will lie in the interval t. If $f(x_l)$ can be assumed linear over another small interval $\pm e_l$ centred on x_l then the probability that the second individual will match with the first, in the sense that the measurement of attribute l for the second individual lies within $\pm e_l$ of x_l, is $f(x_l)2e_l$. Thus, the probability PM_l that two individuals selected at random will match is

$$PM_l = 2e_l \int \{f(x_l)\}^2 dx_l.$$

This expression is a version of the third law of probability applied to continuous measurements. It is the product of the probability that the first measurement lies within a certain interval and the probability that the second measurement lies within a certain interval conditional on the first lying within that interval. This product is then integrated over all possible values of the measurement. It will be a reasonable approximation to the true probability of a match provided, as stated above, e_i is sufficiently small that $f(x_i)$ can be assumed reasonably linear over the interval $\pm e_i$. The discriminatory power can then be derived as before as

$$DP = 1 - \prod_{l=1}^{p} PM_l.$$

Correlated Normally distributed attributes

The probability density function $f(\mathbf{x})$ for a p-dimensional Normally distributed set of correlated attributes $\mathbf{x} = (x_1, \ldots, x_p)$, where, without loss of generality, the mean can be taken as the origin $(0, \ldots, 0)$, may be expressed as

$$f(\mathbf{x}) = \frac{\sqrt{|\Omega|}}{(2\pi)^{p/2}} \exp\left\{ -\frac{1}{2} \sum_{i,j=1}^{p} \Omega_{ij} x_i x_j \right\}$$

(Smalldon and Moffat, 1973; Anderson, 1984).

The matrix $\Omega = \{\Omega_{ij}; \ i, j = 1, \ldots, p.\}$ is the inverse Σ^{-1} of the variance–covariance matrix Σ. From Section 3.4.2, the $\{i, j\}$ member of Σ is $\sigma_i \sigma_j \rho_{ij}$ where σ_i and σ_j are the standard deviations associated with attributes i and j, respectively, and ρ_{ij} is the correlation coefficient between attributes i and j $(i, j = 1, \ldots, p.)$; $|\Omega|$ is the determinant of Ω. Assume the correlation coefficients are not too close to unity and that the p-dimensional surface is reasonably linear on the volume of dimensions $\pm e_l$ $(l = 1, \ldots, p.)$. Then the probability of a match, generalising the result for uncorrelated attributes can be shown to be

$$PM = \prod_{l=1}^{p} 2e_l \int \cdots \int \{f(x_l)\}^2 \prod_{l=1}^{p} dx_l$$

$$= \frac{\sqrt{|\Omega|}}{\pi^{p/2}} \left\{ \prod_{l=1}^{p} e_l \right\}.$$

Table 4.5   The calculation of discriminating power (*DP*) for Normal distributions of p dimensions

p	DP
1	$1 - e_1/(\pi^{1/2}\sigma_1)$
2	$1 - e_1 e_2/\{\pi\sigma_1\sigma_2\sqrt{(1-\rho_{12}^2)}\}$
3	$1 - e_1 e_2 e_3/\{\pi^{3/2}\sigma_1\sigma_2\sigma_3\sqrt{(1-\rho_{12}^2-\rho_{13}^2-\rho_{23}^2+2\rho_{12}\rho_{13}\rho_{23})}\}$

and

$$DP = 1 - PM.$$

The explicit forms of this equation for $p = 1$, 2, 3 are given in Table 4.5.

4.5 SIGNIFICANCE PROBABILITIES

4.5.1 Calculation of significance probabilities

During the course of a crime a window has been broken. A suspect is apprehended soon afterwards and a fragment of glass found on his clothing. Denote this fragment by F. It is of interest to assess the uncertainty relating to whether the fragment came from the broken window. The assessment will be discussed for the moment in the context of an approach based on what are known as *significance probabilities*. Later, in Section 4.6 and Section 7.4.2, two other approaches to this assessment problem, based on coincidence probabilities and likelihood ratios, respectively, will be discussed.

Let θ_0 be the value of the parameter representing the refractive index of the broken window. This is assumed constant. In Section 4.6 and Section 7.4.2 this assumption will be dropped and θ_0 will be replaced by a set of sample measurements of the refractive index from the broken window.

Let x be the refractive index for F. This measurement may be considered as an observation of a random variable X, there being variation in refractive index within a particular window. It is assumed that, if F came from a window of refractive index θ, then X is such that

$$X \sim N(\theta, \sigma^2),$$

(Section 3.4.1). The question at issue is whether F came from the window at the crime scene (and, by association, that the suspect was present at the crime scene). If this is so then θ will be equal to θ_0.

An argument based on significance probabilities is as follows. Suppose $\theta = \theta_0$. The further inference that F came from the window at the crime scene requires the assumption that the mean refractive index is unique to that window; this is not a particularly statistical assumption and is something that should perhaps be part of I, the background information.

The supposition that $\theta = \theta_0$ will be referred to as the *null hypothesis* and denoted H_0. Other names for such a hypothesis are *working hypothesis* or *status quo*. This nomenclature is not particularly appropriate here. It does not seem reasonable to start the analysis with the hypothesis that the suspect was at the scene of the crime. Nonetheless, this line of reasoning is pursued as the statistical ideas on which it is based are in very common usage.

Under the supposition that $\theta = \theta_0$, the deviation (in absolute terms, independent of sign) of x, the refractive index of F, from θ_0 would be expected to be small—just what is meant by small depends on σ, the standard deviation. The distribution of X has been taken to be Normal. The deviation of an observation x from the mean θ of the distribution is measured in terms of the probability of observing a value for the random variable X as extreme as x. This probability cannot be determined analytically and reference has to be made to tables of probabilities of the standard Normal distribution (e.g., Lindley and Scott, 1984).

Let Z be a random variable with a standard Normal distribution; thus

$$Z \sim N(0, 1).$$

The probability that Z is less than a particular value z, $Pr(Z < z)$, is denoted $\Phi(z)$. Certain values of z are used commonly in the discussion of significance probabilities, particularly those values for which $1 - \Phi(z)$ is small, and some of these are tabulated in Table 4.6. Corresponding probabilities for absolute values of Z may be deduced from the tables by use of the symmetry of the Normal distribution. By symmetry,

$$\Phi(-z) = Pr(Z < -z) = 1 - Pr(Z < z) = 1 - \Phi(z).$$

Thus

$$Pr(|Z| < z) = Pr(-z < Z < z)$$
$$= Pr(Z < z) - Pr(Z < -z)$$
$$= \Phi(z) - \Phi(-z)$$
$$= 2\Phi(z) - 1.$$

Particular, commonly used, values of z with the corresponding probabilities for the absolute values of z are given in Table 4.7.

Table 4.6 Values of cumulative distribution function $\Phi(z)$ and its complement $1 - \Phi(z)$ for the standard Normal distribution for given values of z

z	$\Phi(z)$	$1 - \Phi(z)$
1.6449	0.950	0.050
1.9600	0.975	0.025
2.3263	0.990	0.010
2.5758	0.995	0.005

Table 4.7   Probabilities for absolute values from the standard
Normal distribution function

z	$\Phi(z)$	$Pr(|Z| < z) = 2\Phi(z) - 1$	$Pr(|Z| > z)$
1.6449	0.950	0.90	0.10
1.9600	0.975	0.95	0.05
2.3263	0.990	0.98	0.02
2.5758	0.995	0.99	0.01

Figure 4.1 illustrates the probabilities for the following events:

(a) $Pr(Z > 1) = 0.159$,

(b) $Pr(Z > 2) = 0.023$,

(c) $Pr(|Z| < 2) = Pr(-2 < Z < 2) = 0.954$,

(d) $Pr(Z > 3) = 0.0015$.

Consider again the glass problem. If H_0 is true,

$$X \sim N(\theta_0, \sigma^2).$$

Let $Z = (X - \theta_0)/\sigma$. Then

$$Z \sim N(0, 1).$$

Also, $Pr(|X| > x) = Pr(|Z| > z)$.

(a)

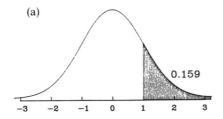

(b)

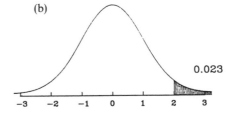

(c)

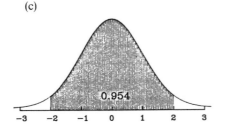

(d)

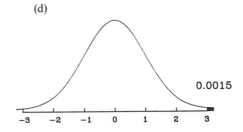

Figure 4.1   Probabilities for various values of a standard Normal random variable (a)
$Pr(Z > 1)$ (b) $Pr(Z > 2)$ (c) $Pr(|Z| < 2)$ (d) $Pr(Z > 3)$.

The probability $Pr(|X| > x)$ is the probability of what has been observed (x) or anything more extreme if H_0 ($\theta = \theta_0$) is true (and hence, as discussed above, that F came from the window at the crime scene). The phrase 'anything more extreme' is taken to mean anything more extreme in relationship to an implicit alternative hypothesis that if H_0 is not true then $\theta \neq \theta_0$. The distance of an observation x from a mean θ is measured in terms of the standard deviation σ. For example, if $\theta = 1.518\,458$ and $\sigma = 4 \times 10^{-5}$, a value of x of $1.518\,538$ is $(1.518\,538 - 1.518\,458)/4 \times 10^{-5}$, or 2.0, standard deviations from the mean. A value of the refractive index x which is more extreme than $1.518\,538$ is one which is more than two standard deviations from the mean in *either* direction, i.e., a value of x which is greater than $1.518\,538$ or less than $1.518\,378$.

The probability, of what is observed or anything more extreme, calculated assuming the null hypothesis is true, is known as the *significance probability*. It may be thought to provide a measure of compatibility of the data with the null hypothesis. It is conventionally denoted P. A small value of P casts doubt on the null hypothesis. In the example discussed here, a small value would cast doubt on the hypothesis that F came from the window at the crime scene. However, it is not clear what value to take for 'small'.

Certain values of P have been used to provide values known as *significance levels* at which a scientist decides to act as if the null hypothesis is false. Typical values are 0.10, 0.05 and 0.01. Thus, for example, a value x of the refractive index for which $P < 0.05$ would be said to be *significant at the 5% level*. If an approach to the evaluation of evidence based on significance levels is taken then the choice of what level of P to choose is obviously of crucial importance. It is helpful when deciding on a level for P, to bear in mind the implications of the decision. The significance probability P is the probability of what is observed or anything more extreme if the null hypothesis is true. Suppose a significance level of 0.05 has been chosen. Then, by chance alone, on 5% of occasions on which this test is conducted, a decision will be made to act as if the null hypothesis is false and a wrong decision will have been made.

Consider the glass example again. On 5% of occasions, using a significance level of 5%, in which such a test was conducted and in which F came from the window at the crime scene it will be decided to act as if F did not come from the window. This decision will be wrong. Obviously there will be many other factors which affect the decision in any particular case. However, the principle remains. The use of this type of analysis gives rise to the probability of an error. It is a well-known error and is given the name *type 1 error or error of the first kind*.

The probability of a type 1 error can be reduced by reducing the significance level, e.g., to 0.01 or 0.001. However, this can only be done at the expense of increasing the probability of another type of error, known as a *type 2 error or error of the second kind*. A type 2 error is the error of failing to reject the null hypothesis when it is false. In the example, it will be the error of deciding that F did come from the window at the crime scene when in fact it did not. In general, if other factors, such as the number of fragments considered, remain constant, it is not possible to

Table 4.8 Significance probabilities P for refractive index x of glass for mean $\theta_0 = 1.518\,458$ and standard deviation $\sigma = 4 \times 10^{-5}$ and decisions assuming a significance level of 5%

x	$z = (x - \theta_0)/\sigma$	$P = Pr(|X| > x) = Pr(|Z| > z)$	Decide to act as if F did or did not come from the crime window
1.518 500	1.05	0.29	Did
1.518 540	2.05	0.04	Did not
1.518 560	2.55	0.01	Did not

choose a significance level so as to decrease the probabilities of both type 1 and type 2 errors simultaneously.

Assume F came from the window at the scene of the crime. Then $\theta = \theta_0$. As a numerical illustration, let $\theta_0 = 1.518\,458$ and $\sigma = 4 \times 10^{-5}$. Illustrations of the calculations for the significance probability are given in Table 4.8.

Determination of the probability of a type 2 error requires knowledge of the value of θ if F did not come from the window at the crime scene. Normally such knowledge is not available and the probability cannot be determined. There may, occasionally, be circumstances in which the probability of a type 2 error can be determined. For example, this could happen if the defence were to identify another source for F. However, even if a probability cannot be determined, one has to be aware that such an error may be made.

Notice that the philosophy described here fits uncomfortably with the legal philosophy that a person is innocent until proven guilty. The null hypothesis in the example is that F came from the window at the crime scene. Failure to reject this hypothesis is an implicit acceptance that the suspect was at the crime scene. Yet, it is the null hypothesis which is being tested. It is the one which has to be rejected or not. The calculation of the significance probability P is based on the assumption that the null hypothesis is true. Only if P is small (and it is values of 0.05, 0.01 and 0.001 which have been suggested as small) is it considered that the null hypothesis may be false. Evidence is required to show that the suspect has *not* been at the crime scene. The principle that one is innocent until proven guilty requires that evidence be produced to show that the suspect has been at the crime scene, not the opposite.

The interpretation of P has to be made carefully. Consider a result from Table 4.8: $\theta_0 = 1.518\,458$, $\sigma = 4 \times 10^{-5}$, $x = 1.518\,560$ and $P = 0.01$. This may be written more explicitly in the notation of conditional probability as

$$Pr(|X| > x | \theta = \theta_0, \sigma) = 0.01. \tag{4.4}$$

It is difficult to relate this to the matter at issue: was the suspect at the crime scene? A small value would seem indicative of the falseness of the null hypothesis but it is not the probability that the null hypothesis is true. It is incorrect to use the value for P as the probability that the suspect was at the crime scene. The transposed probability

$$Pr(\theta = \theta_0 \mid |X| > x, \sigma) \tag{4.5}$$

would be much more useful but this is not what has been calculated. The relationship between (4.4) and (4.5) is similar to that between the probability of the evidence given the suspect is guilty and the probability that the suspect is guilty given the evidence. The interpretation of the first of the last two probabilities as the second is the fallacy of the transposed conditional (Section 2.3.1). It is possible, however, to discuss the relationship between the significance probability and the probability that the suspect was at the crime scene, through the use of likelihood ratios.

4.5.2   Relationship to likelihood ratio

The relationship between significance probabilities and the likelihood ratio has been investigated by many authors. Early references include Good (1956), Lindley (1957) and Edwards *et al.* (1963). The discussion here in the context of the refractive index of glass is based on Berger and Sellke (1987).

Consider two competing and complementary hypotheses:

- C: the fragment of glass on the suspect's clothing came from the window at the crime scene;
- $\bar{C}$: the fragment of glass on the suspect's clothing did not come from the window at the crime scene.

Let p denote the probability that C is true and $(1 - p)$ the probability that $\bar{C}$ is true. If C is true then θ the mean refractive index of the source of the fragment F on the suspect's clothing is θ_0. If $\bar{C}$ is true, then F is assumed to come from a window whose mean refractive index is not equal to θ_0.

Assume C is true. Denote the probability density function of X, the refractive index of F, by $f(x \mid \theta_0, C)$. In this context, this is a Normal density function.

Assume $\bar{C}$ is true. Denote the probability density function of X by $f(x \mid \bar{C})$. The mean θ of the refractive index of the source of the fragment on the suspect's clothing may also be thought of as a random variable which varies from window to window over some relevant population of windows. As such it also has a probability density function; denote this by $f(\theta)$. If θ is known, the probability density function of x is given by $f(x \mid \theta)$. An extension of the law of total probability (Section 1.6.6) to continuous data with integration replacing summation may be used to give the expression

$$f(x \mid \bar{C}) = \int f(x \mid \theta) f(\theta) \, d\theta.$$

The probability density function of X, independent of C and $\bar{C}$, is then

$$f(x) = f(x|\theta_0, C)p + (1 - p)f(x|\bar{C}).$$

Thus, the probability that C is true, given the measurement x is

$$Pr(C|x) = f(x|\theta_0, C)p/f(x)$$

$$= \left\{1 + \frac{(1 - p)f(x|\bar{C})}{pf(x|\theta_0, C)}\right\}^{-1}, \tag{4.6}$$

an expression similar to one used in paternity cases (see Section 6.8.1).

The posterior odds in favour of C is then, using a version of (2.9),

$$\frac{Pr(C|x)}{1 - Pr(C|x)} = \frac{p}{1 - p} \frac{f(x|\theta_0, C)}{f(x|\bar{C})}$$

where $\{p/(1 - p)\}$ is the prior odds in favour of C and $\{f(x|\theta_0, C)/f(x|\bar{C})\}$ is the likelihood ratio V of (2.10).

As an illustration of the calculation of the likelihood ratio, assume that if θ is not equal to θ_0 it is a random variable which has a Normal distribution with mean θ_0 and variance τ^2, where, typically, $\tau^2 \gg \sigma^2$. Then

$$f(x|\bar{C}) = \int f(x|\theta)f(\theta)\,d\theta$$

and so

$$(X|\bar{C}) \sim N(\theta_0, \sigma^2 + \tau^2).$$

(See Section 7.2.2 for a fuller derivation of this result.) With $\tau^2 \gg \sigma^2$, the distribution of $(X|\bar{C})$ is approximately $N(\theta_0, \tau^2)$. The likelihood ratio is, thus,

$$V = \frac{f(x|\theta_0, C)}{f(x|\bar{C})} = \frac{(2\pi\sigma^2)^{-1/2}\exp\{-(x - \theta_0)^2/2\sigma^2\}}{(2\pi\tau^2)^{-1/2}\exp\{-(x - \theta_0)^2/2\tau^2\}}.$$

Consider $\tau = 100\sigma$. Let $z^2 = (x - \theta_0)^2/\sigma^2$ be the square of the standardised distance between the observation x and the mean specified by the null hypothesis, θ_0. Then

$$V = 100\exp\left(-\frac{z^2}{2} + \frac{z^2}{2 \times 10^4}\right)$$

$$\simeq 100\exp(-z^2/2).$$

As an illustration, consider $x = 1.518\,540$, $\theta_0 = 1.518\,458$, $\sigma = 4 \times 10^{-5}$, $\tau = 4 \times 10^{-3}$. Then $z^2 = 2.05^2$ and $P = 0.04$, as before (see Table 4.8) which, at

Table 4.9 Variation in the likelihood ratio V, as given by (4.7), with sample size n, for a standardised distance $z_n = 2$, a result which is significant at the 5% level

n	V
1	14
5	30
10	43
20	61

the 5% level, would lead to rejection of the hypothesis that the fragment of glass came from the window at the scene of the crime. However,

$$V = 100 \exp(-2.05^2/2)$$

$$= 12.2,$$

a value for V which, on the verbal scale in Table 2.9, increases the support for C against $\bar{C}$. Such an apparent contradiction is not a new idea and has been graced by the name of Lindley's paradox (see, for example, Good, 1956; Lindley, 1957; Edwards *et al.*, 1963; and Lindley, 1980, for a reference by Lindley to 'Jeffrey's paradox'.)

Suppose that several, $\cdot n \cdot$ say, fragments of glass were found on the suspect's clothing rather than just one. Let $\bar{x}$ denote the mean of these fragments. Then,

$$(\bar{X}|\theta) \sim N(\theta, \sigma^2/n).$$

If $\bar{C}$ is true

$$\bar{X} \sim N(\theta_0, \tau^2 + \sigma^2/n).$$

The likelihood ratio is

$$V = \frac{(2\pi\sigma^2/n)^{-1/2} \exp\{-n(\bar{x} - \theta_0)^2/2\sigma^2\}}{\{2\pi(\tau^2 + \sigma^2/n)\}^{-1/2} \exp[-(\bar{x} - \theta_0)^2/\{2(\tau^2 + \sigma^2/n)\}]}$$

$$\simeq \frac{\tau\sqrt{n}}{\sigma} \exp\left(-\frac{(\bar{x} - \theta_0)^2}{2}\left\{\frac{n}{\sigma^2} - \frac{1}{\tau^2}\right\}\right)$$

$$\simeq 100\sqrt{n} \exp\left\{-\frac{n(\bar{x} - \theta_0)^2}{2\sigma^2}\right\}.$$

The square of the standardised distance, z_n, between $\bar{x}$ and θ_0 is

$$z_n^2 = n(\bar{x} - \theta_0)^2/\sigma^2$$

and thus

$$V = 100\sqrt{n}\exp(-z_n^2/2),\tag{4.7}$$

a value which increases in direct proportion to the square root of the sample size. Suppose $z_n = 2$, a value which is significant at the 5% level in a test of the hypothesis $\theta = \theta_0$ against the alternative hypothesis $\theta \neq \theta_0$. The value of V for various values of n is given in Table 4.9. In each case, a result which is significant at the 5% level has a likelihood ratio which lends support to the hypothesis that the fragments of glass found on the suspect came from the window at the crime scene.

4.5.3   Combination of significance probabilities

Significance probabilities also combine in a different way from the probabilities of events. From the third law of probability (1.6), the product of two probabilities, for dependent or independent events, will never be greater than either of the individual components of the product. However, it is possible, for characteristics which are dependent, that the significance probability of the joint observation may be greater than either of the individual significance probabilities.

Suppose, for example, that as well as the refractive index of the glass in the current example that the density has also been measured. The suffices 1 and 2 are introduced to denote refractive index and density, respectively. Then $\mathbf{x} = (x_1, x_2)'$ is a vector which denotes the measurements of the two characteristics of the glass found on the suspect's clothing and $\boldsymbol{\theta} = (\theta_1, \theta_2)'$ is a vector which denotes the mean values of refractive index and density of glass from the window at the crime scene.

Let $\theta_1 = 1.518\,458$ (the original θ_0 above), $x_1 = 1.518\,540$ and $\sigma_1 = 4 \times 10^{-5}$. Then, the significance probability (P_1, say) for the refractive index of F is 0.04 (see Table 4.8). Suppose $\theta_2 = 2.515\,\mathrm{g\,cm}^{-3}$ with standard deviation $\sigma_2 = 3 \times 10^{-4}\mathrm{g\,cm}^{-3}$ and let $x_2 = 2.515\,615\,\mathrm{g\,cm}^{-3}$ be the density measured on F. The standardised statistic, z_2 say, is then

$$z_2 = (2.515\,615 - 2.515)/0.0003 = 2.05$$

and the significance probability (P_2, say) for the density measurement is also 0.04.

The product of P_1 and P_2 is 0.0016. However, this is not the overall significance probability. The correlation between the refractive index and the density has to be considered.

Let the correlation coefficient, ρ say, between refractive index and density be 0.93 (Dabbs and Pearson, 1972) and assume that the joint probability density function of refractive index and density is bivariate Normal (3.6) with mean θ and covariance matrix Σ. For a bivariate Normal variable $\mathbf{X}$ it can be shown that the statistic U, given by

$$U = (\mathbf{X} - \theta)' \Sigma^{-1} (\mathbf{X} - \theta)$$

has a chi-squared distribution with two degrees of freedom:

$$U \sim \chi_2^2$$

(Mardia *et al.* 1979, p. 39). Values of the χ^2 distribution are well tabulated, see, for example, Lindley and Scott (1984).

The overall significance probability for the two characteristics can be determined by calculating U and referring the answer to the χ^2 tables. The covariance matrix Σ is given by (3.7)

$$\Sigma = \begin{pmatrix} \sigma_1^2 & \rho\sigma_1\sigma_2 \\ \rho\sigma_1\sigma_2 & \sigma_2^2 \end{pmatrix}$$

$$= \begin{pmatrix} (4 \times 10^{-5})^2 & 1.116 \times 10^{-8} \\ 1.116 \times 10^{-8} & (3 \times 10^{-4})^2 \end{pmatrix}$$

$$= \begin{pmatrix} 1.6 \times 10^{-9} & 1.116 \times 10^{-8} \\ 1.116 \times 10^{-8} & 9 \times 10^{-8} \end{pmatrix}.$$

The deviation of the observation $\mathbf{x}$ from the mean θ is $(\mathbf{x} - \theta)' = (8.2 \times 10^{-5}, 6.15 \times 10^{-4})'$. Some rather tedious arithmetic gives the result that $(\mathbf{x} - \theta)' \Sigma^{-1} (\mathbf{x} - \theta) = 4.204$, a result which has a significance probability $P = 0.1225$. Each individual characteristic is conventionally significant at the 5% level yet together the result is not significant at the 10% level.

Considerable care has to be taken in the interpretation of significance probabilities. As a measure of the value of the evidence their interpretation is difficult.

4.6 COINCIDENCE PROBABILITIES

4.6.1 Introduction

One of the criticisms which was levelled at the probabilistic evidence in the Collins' case was the lack of justification for the relative frequency figures which were quoted in court. The works of Tippett *et al.* (1968) and Gaudette and Keeping (1974) were early attempts to collect data. The lack of data relating to the distribution of measurements on a characteristic of interest is a problem which still exists and is a major one in forensic science. If measurements of certain characteristics (such as glass fragments and their refractive indices) of evidence left at the scene of a crime is found to be similar to evidence (measurements of the same characteristics) found on a suspect, then the forensic scientist wants to know

(a) how similar are they, and
(b) is this similarity one that exists between rare characteristics or common characteristics?

In certain cases, such data do exist which can help to answer these questions.

The most obvious example of this is blood grouping. There are various systems of blood grouping and the relative population frequencies of the various categories in each of the systems are well tabulated for different populations (see, for example, Gaensslen *et al.*, 1987a, b, c). Thus if a blood stain of a particular group is found at the scene of a crime there are several possibilities:

(a) it came from the victim,
(b) it came from the criminal,
(c) it came from some third party.

If the first and third possibilities can be eliminated and the blood group is only found in $x\%$ of the population then the implication is that the criminal belongs to that $x\%$ of the population; *i.e.*, he belongs to a subset of the population with $Nx/100$ members, where N is the total population size and $Nx/100$ is of necessity no less than 1. At its simplest this implies that, *all other things being equal*, a person with this blood group has probability $100/Nx$ of being the criminal. This is a consequence of the defender's fallacy explained in Section 2.3.4.

Another example in which data exist, though not to the extent to which they do for blood grouping, is the refractive index of glass particles. Data relating to the refractive index have been collected over several years by the U.K. Forensic Science Laboratories in the course of case work (see, for example, Lambert and Evett, 1984). From a statistical point of view this approach may lead to a biased sample since only glass fragments which have been connected with crimes will be collected. However, as an example of the recognition of this problem a survey of glass found on footpaths has been conducted (Walsh and Buckleton, 1986) to obtain data unrelated with crimes.

An approach based on probabilities, known as *coincidence probabilities*, with relation to the refractive index of glass has been developed by Evett and Lambert in a series of papers (Evett, 1977, 1978; Evett and Lambert, 1982, 1984, 1985), using data from Dabbs and Pearson (1972). Two questions are asked.

(a) Are the source and receptor fragments similar in some sense?
(b) If they are similar are the characteristics rare or common?

A very simple—and trivial—example of this is the following. An eye-witness to a crime says that he saw a person running from the scene of the crime and that this person had two arms. A suspect is found who also has two arms. The suspect is certainly similar to the person running away from the crime but the character-istic—two arms—is so common as not to be worth mentioning. The elementary nature of the example may be extended by noting that if the eyewitness said the criminal was a tall man then any short women, despite having two arms, would be eliminated. In contrast, suppose that the eye-witness says that the person running away had only one arm and a suspect is found with only one arm. This similarity is much stronger evidence that the suspect was at the scene of the crime since the characteristic—one arm—is rare.

A more complicated—and considerably less trivial—application relates to the interpretation of evidence on the refractive index of glass. The coincidence method to be explained is capable of development in this context because of the existence of data for the distribution of measurements of the refractive index of glass. Development is not possible in the examples mentioned earlier of paint fragments and hair comparisons because appropriate data do not exist.

A crime is committed and fragments of glass from a window are found at the scene of the crime; these are the source fragments. A suspect is found and he has fragments of glass on his person; these are the receptor (or transferred-particle) fragments. The interpretation of the data is treated in two stages corresponding to the two questions asked earlier. First, the refractive index measurements are compared by means of a statistical criterion which takes account of between- and within-window variation. Secondly, if the two sets of measurements are found to be similar then the significance of the result is assessed by referring to suitable data.

Various assumptions are needed for these two stages to be applied. Those adopted here are the following.

(a) The refractive index measurements on glass fragments from a broken window have a Normal probability distribution centred on a mean value θ which is characteristic of that window and with a variance σ^2.

(b) The mean θ varies from window to window and in the population of windows θ has its own probability distribution of, as yet, unspecified form.

(c) The variance σ^2 is the same for all windows and it is known.

(d) All the transferred particle fragments are assumed to have come from the same source. A criterion based on the range of the measurements can be devised to check this (e.g., Evett, 1978).

(e) The transferred particle fragments are all window glass.

4.6.2   Comparison stage

The comparison is made by considering the difference, suitably scaled, in the mean of the source fragments measurements and the mean of the receptor fragment measurements. The test statistic is

$$Z = \frac{\bar{X} - \bar{Y}}{\sigma(m^{-1} + n^{-1})^{1/2}}$$

where $\bar{X}$ is the mean of m measurements on source fragments and $\bar{Y}$ is the mean of n measurements on receptor fragments. It is assumed that Z has a standard $N(0, 1)$ distribution. Thus if $|Z| > 1.96$ it is concluded that the receptor fragments are not similar to the source fragments and if $|Z| < 1.96$ it is concluded that the receptor fragments are similar to the souce fragments. The value 1.96 is chosen so that the probability of a type 1 error (deciding that receptor fragments have a different origin from source fragments) is 0.05. Other possible values may be chosen such as 1.65, 2.33, 2.58 which have type 1 error probabilities of 0.10, 0.02 and 0.01, respectively; see Table 4.7.

This statistic and the associated test provide an answer to question (a).

4.6.3   Significance stage

If it is concluded that the receptor and source fragments are not similar, the matter ends there. If it is concluded that they are similar, then there are two possibilities.

(a) The receptor and source fragments came from the same source.
(b) The receptor and source fragments did not come from the same source.

For the assessment of significance the probability of a coincidence (known as a *coincidence probability*) is estimated. This is defined as 'the probability that a set of n fragments taken at random from some window selected at random from the population of windows would be found to be similar to a control window with mean refractive index $\bar{X}$'. It is denoted $C(\bar{X})$. Compare this definition of a probability of coincidence with that given in the section on discriminating power. There the receptor and source fragments were both taken to be random samples from some underlying population and the probability estimated was that of two random samples having similar characteristics. Here the mean $\bar{X}$ of a sample of fragments from the source window is taken to be fixed and the concern is with the estimation of the probability that one sample, namely the receptor fragments, will be found to be similar to this source window. Any variability in the value of $\bar{X}$ is ignored; methods of evaluation which account for this variability are discussed in Chapter 7.

Note that there are two levels of variation which have been considered. First there is the probability that a window selected at random from a population of windows will have a mean refractive index in a certain interval, $(u, u + \mathrm{d}u)$ say; denote this probability by $p(u)$. In practice for this method the data used to estimate $p(u)$ is represented as a histogram and the probability distribution is taken as a discrete distribution over the categories forming the histogram, e.g., $\{p(u_1), \ldots, p(u_k)\}$ where there are k categories. Secondly, the probability is required that n fragments selected at random from a window of mean value u would prove to be similar to fragments from a source window of mean $\bar{X}$; denote this probability by $S_{\bar{x}}(u)$. It is then possible to express $C(\bar{X})$ as a function of $p(u)$ and $S_{\bar{x}}(u)$, namely:

$$C(\bar{X}) = \sum_{i=1}^{k} p(u_i) S_{\bar{x}}(u_i).$$

The derivation of this is given in Evett (1977).

Let $\{x_1, \ldots, x_m\}$, with mean $\bar{x}$, denote the source measurements of refractive index from fragments from a window W broken at the scene of a crime; with

$$\bar{X} \sim N(w, \sigma^2/m),$$

where w is the mean refractive index of window W.

Let $\{y_1, \ldots, y_n\}$, with mean $\bar{y}$, denote the transferred particle measurements of refractive index from fragments T found on a suspect's clothing. If the fragments T have come from W then

$$\bar{Y} \sim N(w, \sigma^2/n).$$

Since the two sets of measurements are independent,

$$(\bar{X} - \bar{Y}) \sim N(0, \sigma^2(m^{-1} + n^{-1})),$$

and, hence,

$$\frac{\bar{X} - \bar{Y}}{\sigma(m^{-1} + n^{-1})^{1/2}} \sim N(0, 1).$$

Denote this statistic by Z. Thus, it can be said that if the fragments T came from W and if the distributional assumptions are correct, the probability that Z has a value greater than 1.96 in absolute terms; i.e.,

$$|Z| > 1.96,$$

is 0.05. Such a result may be thought unlikely under the original assumption that the fragments T have come from W. Hence the original assumption may be questioned. It may be decided to *act as if* the fragments T did not come from W. Evett (1977) showed that with this decision rule at the comparison stage, then the probability $S_{\bar{x}}(u)$ at the significance stage is given by

$$S_{\bar{x}}(u) = \Phi\{(\bar{X} - u)m^{1/2}/\sigma + 1.96(1 + m/n)^{1/2}\}$$
$$- \Phi\{(\bar{X} - u)m^{1/2}/\sigma - 1.96(1 + m/n)^{1/2}\}, \qquad (4.8)$$

from which the coincidence probability $C(\bar{X})$ may be evaluated in any particular case. Some results of the application of this method in comparison with others are given in Section 7.4.2 and Table 7.6.

4.7   LIKELIHOOD RATIO

This is discussed at considerable length in later chapters. The likelihood ratio has been introduced in Section 2.5.1 as the ratio $Pr(E|G)/Pr(E|\bar{G})$ which converts prior odds in favour of guilt (G) into posterior odds in favour of guilt, given evidence E. However, an interesting and somewhat prophetic comment was made by Parker and Holford (1968), in which the comparison problem was treated as in two stages (similarity and discrimination) rather than in one, as an analysis using the likelihood ratio does. The two-stage approach follows, they said, the traditions of forensic scientists. Interestingly, they then went on to make the following remark.

> 'We could (therefore) set up an index R whose numerator is the likelihood that the crime hair comes from the suspect and whose denominator is the likelihood that it comes from the population at large. In ordinary language one would then assert that it was R times more likely for the hair to have come from the suspect than from someone drawn at random from the population. But a statement of this nature is of rather limited use to forensic scientists, by-passing as it does the similarity question altogether.'
> (Parker and Holford, 1968.)

It will be explained in later chapters how the use of the likelihood ratio, which is all that Parker and Holford's index R is, has rather more than 'limited use' and how its use considers similarity in a natural way.

Transfer Evidence

5.1 THE LIKELIHOOD RATIO

5.1.1 Probability of guilt

Uncertainty may be measured by probability. However, attempts to measure uncertainty in forensic evidence by probability alone lead to difficulties in interpretation. The two-stage approach described in Chapter 4 illustrates this. Also, as discussed in Section 2.3.1, there is considerable potential for confusion between the following two conditional probabilities; the probability of the evidence given the guilt of the suspect and the probability of the guilt of the suspect given the evidence. Attempts to evaluate combinations of different pieces of evidence are also fraught with difficulty, requiring consideration of the dependence of these different pieces; see Cohen (1977, 1988) and Dawid (1987) for a debate about this, a debate which is also summarised below.

The scientist is concerned with the uncertainty regarding his evidence. He is not concerned with the guilt or otherwise of the suspect. This is the concern of the court. The court has to take account of much other information and evidence, of which the scientist is not aware, presented to it.

Consider the following example of the confusion that may arise if probabilities of guilt are determined for two pieces of evidence E_1 and E_2. For both E_1 and E_2, suppose the court determines $Pr(G|E_1)$ and $Pr(G|E_2)$ to be 0.7. On a balance of probabilities, E_1 and E_2, separately, imply the guilt of the suspect. The court then multiplies these together, a multiplication which assumes some sort of independence, to produce a probability of $0.7^2 = 0.49$. This last probability is less than 0.5. Superficially, this seems to imply that two pieces of evidence, which separately imply the guilt of the suspect, when combined imply the innocence of the suspect. This apparent contradiction is part of the basis of the criticism by Cohen (1977) of the relevance of the calculations of standard probability (Cohen's term is 'Pascalian'). However, Dawid (1987) explained how a rigorous Bayesian analysis is able to counter ths criticism.

The posterior probability of guilt depends on the prior probability of guilt. Consider the two pieces of evidence above, such that $Pr(G|E_1) = Pr(G|E_2) = 0.7$. The temptation, succumbed to above, which suggests that these two separate probability statements when combined imply that $Pr(G \mid E_1, E_2) = 0.49$, should be resisted. The odds form of Bayes' Theorem (Section 2.4) clarifies the position. The version of the Theorem which is applicable here states that

$$\frac{Pr(G \mid E_1, E_2)}{Pr(\bar{G} \mid E_1, E_2)} = \frac{Pr(E_1, E_2 \mid G)}{Pr(E_1, E_2 \mid \bar{G})} \times \frac{Pr(G)}{Pr(\bar{G})}.$$

A clear understanding is needed of the meaning of 'independence' when applied to the two pieces of evidence, E_1 and E_2. In this context independence means that the joint probability of the two pieces of evidence, given guilt (G) or innocence ($\bar{G}$) is equal to the product of the individual probabilities. Thus

$$Pr(E_1, E_2 \mid G) = Pr(E_1 \mid G) \times Pr(E_2 \mid G)$$

and

$$Pr(E_1, E \mid \bar{G}) = Pr(E_1 \mid \bar{G}) \times Pr(E_2 \mid \bar{G}).$$

This does not mean, for example, that $Pr(G \mid E_1, E_2) = Pr(G \mid E_1) \times Pr(G \mid E_2)$. Then

$$\frac{Pr(E_1, E_2 \mid G)Pr(G)}{Pr(E_1, E_2 \mid \bar{G})Pr(\bar{G})} = \frac{Pr(E_1 \mid G)Pr(E_2 \mid G)Pr(G)}{Pr(E_1 \mid \bar{G})Pr(E_2 \mid \bar{G})Pr(\bar{G})}.$$

If $Pr(G) = Pr(\bar{G})$ this equals

$$\frac{Pr(E_1 \mid G)Pr(G)}{Pr(E_1 \mid \bar{G})Pr(\bar{G})} \times \frac{Pr(E_2 \mid G)Pr(G)}{Pr(E_2 \mid \bar{G})Pr(\bar{G})}$$

$$= \frac{Pr(G \mid E_1)}{Pr(\bar{G} \mid E_1)} \times \frac{Pr(G \mid E_2)}{Pr(\bar{G} \mid E_2)}$$

$$= \frac{0.7}{0.3} \times \frac{0.7}{0.3}$$

$$= \frac{0.49}{0.09}.$$

Thus

$$Pr(G \mid E_1, E_2) = \frac{0.49/0.09}{1 + 0.49/0.09} = \frac{0.49}{0.58} = 0.84 > 0.7.$$

Hence, if the prior odds are equal and if the two pieces of evidence are independent, conditional on the hypotheses of guilt or innocence, then the conjunction of E_1 and E_2 is such as to strengthen the probability of guilt.

5.1.2 Justification

The odds form of Bayes' Theorem presents a compelling intuitive argument for the use of the likelihood ratio as a measure of the value of the evidence. A mathematical argument exists also to justify its use. A simple proof is given in Good (1991) and repeated here for convenience.

It is desired to measure the value V of evidence E in favour of guilt G. There will be dependence on background information I but this will not be stated explicitly. It is assumed that this value V is a function only of the probability of E given that the suspect is guilty, G, and of the probability of E given that the suspect is innocent, $\bar{G}$.

Let $x = Pr(E \mid G)$ and $y = Pr(E \mid \bar{G})$. The assumption above states that

$$V = f(x, y)$$

for some function f.

Consider another piece of evidence T which is irrelevant to (independent of) E and G (and hence $\bar{G}$) and which is such that $Pr(T) = \theta$. Then

$$
\begin{aligned}
Pr(E, T \mid G) &= Pr(E \mid G)Pr(T \mid G) && \text{by the independence of } E \text{ and } T \\
&= Pr(E \mid G)Pr(T) && \text{by the independence of } T \text{ and } G \\
&= \theta x.
\end{aligned}
$$

Similarly,

$$Pr(E, T \mid \bar{G}) = \theta y.$$

The value of the combined evidence (E, T) is equal to the value of E, since T has been assumed irrelevant. The value of (E, T) is $f(\theta x, \theta y)$ and the value of $E = V = f(x, y)$. Thus

$$f(\theta x, \theta y) = f(x, y)$$

for all θ in the interval $[0, 1]$ of possible values of $Pr(T)$. It follows that f is a function of x/y alone and hence that V is a function of

$$Pr(E \mid G)/Pr(E \mid \bar{G}),$$

namely, the likelihood ratio. For most of this book, the likelihood ratio itself is considered as the value of the evidence. However, as explained in Section 2.4.2, additivity in the weighing of evidence is a desirable concept. This can be achieved by taking the function to be a logarithmic one and defining weight of evidence to be the logarithm of the ratio $Pr(E \mid G)/Pr(E \mid \bar{G})$; see Section 2.4.2.

5.1.3 Combination of evidence and comparison of more than two hypotheses

Notice that the representation of the value of evidence as a likelihood ratio enables successive pieces of evidence to be evaluated sequentially in a much more intuitive and simpler way than can be done with significance probabilities, for example as in Section 4.5. The posterior odds from one piece of evidence, E_1 say, become the prior odds for the following piece of evidence, E_2 say. Thus

$$\frac{Pr(G|E_1)}{Pr(\bar{G}|E_1)} = \frac{Pr(E_1|G)}{Pr(E_1|\bar{G})} \times \frac{Pr(G)}{Pr(\bar{G})}$$

and

$$\frac{Pr(G|E_1, E_2)}{Pr(\bar{G}|E_1, E_2)} = \frac{Pr(E_2|G, E_1)}{Pr(E_2|\bar{G}, E_1)} \times \frac{Pr(G|E_1)}{Pr(\bar{G}|E_1)}$$

$$= \frac{Pr(E_2|G, E_1)}{Pr(E_2|\bar{G}, E_1)} \times \frac{Pr(E_1|G)}{Pr(E_1|\bar{G})} \times \frac{Pr(G)}{Pr(\bar{G})}. \tag{5.1}$$

Thus the possible dependence of E_2 on E_1 is recognised in the form of the probability statements within the likelihood ratio.

If the two pieces of evidence are independent (as may be the case with genetic marker systems) this leads to the likelihood ratios combining by simple multiplication:

$$\frac{Pr(E_1, E_2|G)}{Pr(E_1, E_2|\bar{G})} = \frac{Pr(E_1|G)}{Pr(E_1|\bar{G})} \times \frac{Pr(E_2|G)}{Pr(E_2|\bar{G})}.$$

Thus, if V_{12} is the likelihood ratio for the combination of evidence (E_1, E_2), and if V_1 and V_2 are the likelihood ratios for E_1 and E_2, respectively, then

$$V_{12} = V_1 \times V_2.$$

If the weight of evidence is used (Section 2.4.2), different pieces of evidence may be combined by addition. This is a procedure which has an intuitive analogy with the scales of justice.

Example 5.1 · Consider the case of *State v. Klindt* (District Court of Scott County, Iowa, Case Number 115422, 1968) discussed in Lenth (1986). This case has two aspects of interest. It illustrates a method for combining evidence and it illustrates a method for weighing the evidence for more than two hypotheses.

The case involved the identity of a portion of a woman's body. The portion was analysed and it was determined that the woman was white, aged between 27 and

40, had given birth to at least one child and had not been surgically sterilised. Also, seven genetic markers were identified. All but four missing persons in the area of the four states round where the body was found were eliminated as possible identities for the woman. Label these four persons as *P*, *Q*, *R* and *S*. Women *Q*, *R* and *S* had been missing for six months, six years and seven years, respectively, and their last known locations were at least 200 miles from where the body was found. Woman *P* had been missing for one month at the time the body was discovered and had last been seen in the same area.

The blood type of *P* was known to be *A*. The blood types of *Q*, *R* and *S* were not known. The other markers were unknown for all four women. For the remaining six phenotypes, samples of tissues from the parents of *P* enabled a value of 0.5 to be calculated for the probability that the woman whose body was found had the phenotypes it had if it were that of *P*; i.e. $Pr($given phenotypes $\mid P)$ equals 0.5. See also Section 6.7 for another example of this so-called parentage testing. For some general population the incidence of these phenotypes was 0.007 64. No familial testing was done for *Q*, *R* or *S*. The ages of all four women were known and they were all mothers. However, in order to illustrate the procedure for combining evidence, Lenth (1986) made the following alterations to the actual data: it was not known whether *Q* was a mother; *R*'s age was taken to be unknown and *S* was known to have type *A* blood.

The above information is summarised in Table 5.1 using frequency information given by Lenth (1986). If a particular characteristic is known to hold for a particular woman then a probability of 1(corresponding to certainty) is entered.

Table 5.1 Probability evidence in *State v Klindt* (altered for illustrative purposes) from Lenth (1986)

Indicator	Attribute	Woman, X			
		P	Q	R	S
	Age (years)	33	27	unknown	37
E_1	Mother $\mid$ age	1.0	0.583	1.0	1.0
E_2	Not sterilised $\mid$ mother, age	1.0	0.839	0.662	0.542
E_3	Type A blood	1.0	0.362	0.362	1.0
E_4	Other six phenotypes	0.500	0.00764	0.00764	0.00764
$Pr(E_1, E_2, E_3, E_4 \mid$ age, $X)$		0.5000	0.00135	0.00183	0.00414
Fraction of total, $Pr(X \mid, E$ age)		0.9856	0.0027	0.0036	0.0082

If the presence or absence of an attribute is not known, the incidence probability of the attribute amongst the general population is given. There are four hypotheses to be compared, one for each of the four women whose body may have been found. The evidence to be assessed is $E = \{E_1, \ldots, E_4\}$, the four separate pieces of evidence listed in Table 5.1. The probabilities to be determined are of the form $Pr(E_1 \ldots E_4 \mid \text{age}, X)$ where X is one of P, Q, R or S. These pieces of evidence are not all mutually independent. Denote the evidence as follows:

- E_1: mother, yes or no;
- E_2: not sterilised, yes or no;
- E_3: type A blood, yes or no;
- E_4: other six phenotypes.

Using the third law of probability for dependent events (1.7)

$$Pr(E \mid \text{age}, X) = Pr(E_1, E_2, E_3, E_4 \mid \text{age}, X)$$
$$= Pr(E_1 \mid \text{age}, X)Pr(E_2 \mid E_1, \text{age}, X)Pr(E_3 \mid E_2, E_1, \text{age}, X)Pr(E_4 \mid E_3, E_2, E_1, \text{age}, X)$$
$$= Pr(E_1 \mid \text{age}, X)Pr(E_2 \mid E_1, \text{age}, X)Pr(E_3 \mid X)Pr(E_4 \mid X)$$

where X is one of P, Q, R or S, and the final expression depends on certain independence relationships amongst the four pieces of evidence. For example, the probability of sterilisation is dependent on age and being a mother or not, whilst the probability an individual has type A blood (E_3) is independent of age, whether a mother or not (E_1), and whether or not sterilised (E_2). The probabilities for the combined evidence E are given in the penultimate row of Table 5.1, namely

- $Pr(E \mid \text{age}, P) = 0.5000$;
- $Pr(E \mid \text{age}, Q) = 0.001\ 35$;
- $Pr(E \mid \text{age}, R) = 0.001\ 83$;
- $Pr(E \mid \text{age}, S) = 0.004\ 14$.

Explicit dependence on 'age' is now omitted for ease of notation. The posterior probability that the body is that of P, $Pr(P \mid E)$, may be determined so long as information concerning the identity of the body prior to the discovery of the evidence contributing to E is available and is represented by the prior probabilities $Pr(P)$, $Pr(Q)$, $Pr(R)$ and $Pr(S)$. From Bayes' Theorem (2.2)

$$Pr(P \mid E) = \frac{Pr(E \mid P)Pr(P)}{Pr(E)}$$

with similar results for Q, R and S. From the law of total probability (1.9), since the events P, Q, R and S are mutually exclusive and exhaustive

$$Pr(E) = Pr(E \mid P)Pr(P) + Pr(E \mid Q)Pr(Q) + Pr(E \mid R)Pr(R) + Pr(E \mid S)Pr(S).$$

If the four prior probabilities are assumed mutually exclusive, exhaustive and equal,

$$Pr(P) = Pr(Q) = Pr(R) = Pr(S) = 1/4.$$

Hence,

$$Pr(P \mid E) = \frac{Pr(E \mid P)}{Pr(E \mid P) + Pr(E \mid Q) + Pr(E \mid R) + Pr(E \mid S)}.$$

From Table 5.1

$$Pr(P \mid E) = \frac{0.5000}{0.5000 + 0.001\,35 + 0.001\,83 + 0.004\,14} = 0.9856,$$

the value given in the final row of the table. The posterior odds in favour of P then equal $0.5000/(0.001\,35 + 0.001\,83 + 0.004\,14) \simeq 68$. This can be verified by determining $Pr(P \mid E)/Pr(\bar{P} \mid E) = 0.9856/0.0144 \simeq 68$ where $\bar{P}$ is the complement of P, which, for the moment is taken to be Q, R or S.

The value V of the evidence which equals $Pr(E \mid P)/Pr(E \mid \bar{P})$, may be determined, again if the prior probabilities are available. From Bayes' Theorem (2.2)

$$Pr(E \mid \bar{P}) = \frac{Pr(\bar{P} \mid E)Pr(E)}{Pr(\bar{P})}$$

$$= \frac{\{Pr(Q \mid E) + Pr(R \mid E) + Pr(S \mid E)\}Pr(E)}{Pr(Q) + Pr(R) + Pr(S)}$$

$$= \frac{Pr(E \mid Q)Pr(Q) + Pr(E \mid R)Pr(R) + Pr(E \mid S)Pr(S)}{Pr(Q) + Pr(R) + Pr(S)}. \qquad (5.2)$$

If $Pr(Q) = Pr(R) = Pr(S)$ then

$$Pr(E \mid \bar{P}) = \frac{Pr(E \mid Q) + Pr(E \mid R) + Pr(E \mid S)}{3}$$

$$= \frac{0.007\,32}{3}$$

$$= 0.002\,44.$$

Thus

$$\frac{Pr(E \mid P)}{Pr(E \mid \bar{P})} = \frac{0.5000}{0.002\,44} = 204.92.$$

The evidence is about 205 times more likely if the body is that of P than one of the other three women.

If $Pr(P) = Pr(Q) = Pr(R) = Pr(S) = 1/4$ then $Pr(P)/Pr(\bar{P}) = 1/3$ and the posterior odds in favour of P equals

$$\frac{0.5000}{0.002\,44} \times \frac{1}{3} = \frac{0.5000}{0.007\,32} \simeq 68$$

as determined previously.

It is also possible to determine the value of the evidence in favour of P relative to one other woman, Q say, in the usual way of comparing two hypotheses:

$$\frac{Pr(E\,|\,P)}{Pr(E\,|\,Q)} = \frac{0.5000}{0.001\,35} \simeq 370.$$

The evidence is about 370 times more likely if the body is that of P rather than that of Q. As well as age, the conditioning on background information I has, as usual, been omitted for clarity of exposition. However, there is information available regarding the length of time for which the women have been missing and their last known locations. This may be considered as background information I. As such it may be incorporated into the prior probabilities which may then be written as $Pr(P\,|\,I)$, $Pr(Q\,|\,I)$, $Pr(R\,|\,I)$ and $Pr(S\,|\,I)$. Suppose $(P\,|\,I)$ is thought the most likely event and that the other three events are all equally unlikely. Represent this as $Pr(P\,|\,I) = 0.7$, $Pr(Q\,|\,I) = Pr(R\,|\,I) = Pr(S\,|\,I) = 0.1$ (though this is not the only possible combination of probabilities which satisfy this criterion). The likelihood ratio $Pr(E\,|\,P)/Pr(E/\bar{P})$ is the same as before (204.92) since it was determined assuming only that $Pr(Q) = Pr(R) = Pr(S)$, without specifying a particular value. The posterior odds alter, however. The prior odds are

$$\frac{Pr(P\,|\,I)}{Pr(\bar{P}\,|\,I)} = \frac{0.7}{0.3}.$$

The posterior odds then becomes

$$204.92 \times \frac{0.7}{0.3} \simeq 478$$

and $Pr(P\,|\,E) = 0.998$.

The posterior probabilities in Table 5.1 have been calculated assuming $Pr(P) = Pr(Q) = Pr(R) = Pr(S) = 1/4$. If these probabilities are not equal, the posterior probabilities have to be calculated taking account of the relative values of the four individual probabilities. For example,

$$Pr(P\,|\,E) = \frac{Pr(E\,|\,P)Pr(P)}{Pr(E\,|\,P)Pr(P) + Pr(E\,|\,Q)Pr(Q) + Pr(E\,|\,R)Pr(R) + Pr(E\,|\,S)Pr(S)}.$$

The likelihood ratio also has to be calculated again if $Pr(P)$, $Pr(Q)$, $Pr(R)$ and $Pr(S)$ are not equal. From (5.2)

$$\frac{Pr(E\,|\,P)}{Pr(E\,|\,\bar{P})} = \frac{Pr(E\,|\,P)\{Pr(Q) + Pr(R) + Pr(S)\}}{Pr(E\,|\,Q)Pr(Q) + Pr(E\,|\,R)Pr(R) + Pr(E\,|\,S)Pr(S)}. \tag{5.3}$$

These results for comparing four hypotheses may be generalised to any number, n say, of competing exclusive hypotheses. Let $H_1, \ldots, H_n$ be n exclusive hypotheses and let E be the evidence to be evaluated. Denote the probability of E under each of the n hypotheses by $Pr(E\,|\,H_i)$, $(i = 1, \ldots, n)$. Let $p_i = Pr(H_i)$, $(i = 1, \ldots, n)$ be the prior probabilities of the hypotheses. Then consider the value E for comparing H_1 with $(H_2, \ldots, H_n) = \bar{H}_1$, say:

$$\frac{Pr(E\,|\,H_1)}{Pr(E\,|\,\bar{H}_1)} = \frac{Pr(E\,|\,H_1)(\sum_{i=2}^{n} p_i)}{\sum_{i=2}^{n} Pr(E\,|\,H_i)p_i}$$

from a straightforward extension of (5.3). Thus

$$\frac{Pr(E\,|\,H_1)}{Pr(E\,|\,\bar{H}_1)} = \frac{Pr(E\,|\,H_1)(1 - p_1)}{\sum_{i=2}^{n} Pr(E\,|\,H_i)p_i},$$

and

$$Pr(H_1\,|\,E) = \frac{Pr(E\,|\,H_1)p_1}{Pr(E)}$$

$$= \frac{Pr(E\,|\,H_1)p_1}{\sum_{i=1}^{n} Pr(E\,|\,H_i)p_i}.$$

The posterior odds are best evaluated by writing $Pr(\bar{H}_1\,|\,E)$ as $1 - Pr(H_1\,|\,E)$ and then

$$\frac{Pr(H_1\,|\,E)}{Pr(\bar{H}_1\,|\,E)} = \frac{Pr(E\,|\,H_1)p_1/\sum_{i=1}^{n} Pr(E\,|\,H_i)p_i}{1 - \{Pr(E\,|\,H_1)p_1/\sum_{i=1}^{n} Pr(E\,|\,H_i)p_i\}}$$

$$= \frac{Pr(E\,|\,H_1)p_1}{\sum_{i=2}^{n} Pr(E\,|\,H_i)p_i}.$$

Example 5.1 continued—population considerations The probability figures for non-sterilisation of women, and for women who are mothers and of a certain age are based on information about white women in general. It is unlikely that the probabilities are the same among missing white women (Lenth, 1986). However, these results are the best available and serve to illustrate the methodology.

There is a possibility not so far accounted for that the body may be that of a woman other than P, Q, R or S. If such a possibility is to be considered then it can be done by adding an extra hypothesis to the four under consideration and using the general results with $n = 5$. Information is then needed concerning the probability p_5 to be assigned to this hypothesis, remembering to adjust $p_1, \ldots, p_4$ appropriately so that $p_1 + \cdots + p_5 = 1$. Information is also needed concerning $Pr(E \mid H_5)$. Some is available from Table 5.1 by taking relevant information from that pertaining to the other women. Thus $Pr(\text{mother} \mid \text{age unknown} = 0.583$, from Q. However, $Pr(\text{not sterilised} \mid \text{mother unknown, age unknown})$ is unavailable. Probabilities are given in Table 5.1 for mothers of unknown age and for women of a certain age but about whom it is not known if they are mothers or not. A value is not given for the probability that a woman would be sterilised when her age is unknown and it is not known whether she is a mother or not. The probabilities for E_3 (type A blood) and E_4 (other six phenotypes) would remain the same since the unknown woman has been identified from her remains as being white, and the appropriate probabilities have been given.

5.2   CORRESPONDENCE PROBABILITIES

A *correspondence probability* is the probability that evidence related to the suspect corresponds to evidence found at the scene of the crime. It is discussed by Stoney (1991) where it was called a *correspondence frequency*.

A correspondence probability provides an answer to the question

'Given the evidence found at the scene of the crime, what is the probability of encountering corresponding evidence related to the suspect?'

The evidence found at the scene of the crime is accepted. The rarity of a correspondence with this evidence is of interest. Consider Example 1.2 in which the evidence at the scene of the crime is the source or bulk form.

Given the properties of the broken crime scene window, what is the probability of the random occurrence of corresponding glass fragments? The estimation of this probability requires information on a population of glass samples. From this population, the probability of a correspondence with the properties of the window at the scene of the crime can be determined. The frequency of occurrence of this set of properties among glass fragments (the transferred particle form) is the frequency of interest. It is not necessarily the frequency of occurrence amongst windows (the bulk source form).

Extensive data may exist on the properties of window glass but it cannot be assumed that the same frequencies of properties necessarily exist amongst glass fragments found on people. It is not necessarily the case that all glass fragments found on people come from broken windows. They may come from containers or head lamps, for example. The frequencies of properties in the different populations may differ. Rare properties in window glass may be common properties in glass fragments from containers or head lamps.

The results of a search of footpaths in metropolitan Auckland, New Zealand, for pieces of broken, colourless glass was reported by Walsh and Buckleton (1986).

The predominant type of broken glass was that of container glass. The assessment of the probability of a coincidental match of refractive indices can be affected by the type of glass in a survey. A refractive index of 1.5202 has a frequency of 3.1% in container glass in the United Kingdom but a frequency of 0.23% amongst building glass (Lambert and Evett, 1984). Consider a crime in which the refractive index of glass from a broken window at the scene has a refractive index of 1.5202. A suspect may have fragments of container glass of refractive index 1.5202 embedded in his footwear. Reference of this to a database constructed from a survey which is predominantly of building glass will provide a probability of a coincidental match which is too low by a factor greater than 10 (see Section 5.3.3). The value of the evidence against the suspect will be exaggerated. As Walsh and Buckleton (1986) comment, the survey used to assess the significance of glass evidence 'should realistically reflect the type of broken glass likely to be encountered at random in the community'.

The appropriate survey results may depend on where on the suspect the fragments were found. Fragments found on footwear may be expected to have come from a street and have greater proportions of automobile or bottle glass. Fragments found on clothing may be expected to have come from contact with loose fragments on a surface or as a result of being close to a breaking object. In general, population data relevant to the type of data and the environment in which the material associated with the suspect has been found should be used. The survey reported by Walsh and Buckleton (1986) has direct relevance only for the consideration of glass fragments found in footwear. It is not applicable to the consideration of glass fragments found on the clothing of a suspect. A survey of refractive indices of glass recovered from persons suspected of a crime and unrelated to the crime reference samples has been carried out by Harrison *et al* (1985). They reported, unsurprisingly, a predominance of fragments of container glass in footwear but suggested a significant proportion of building glass on clothing.

5.3   DIRECTION OF TRANSFER

It is a feature of likelihood ratios that it is not necessary to distinguish between the scene-anchored and the suspect-anchored perspective; see below and Stoney (1991). However, despite this feature, it is still necessary to recognise the possible importance of the direction of the transfer of the evidence. General results are given in Evett (1984).

5.3.1   Transfer of evidence from the criminal to the scene

Consider Example 1.1 again. A bloodstain is found at the scene of the crime. It is of type Γ. All innocent sources of the stain have been eliminated, the knowledge of

Table 5.2 Frequencies of Ruritanians and those of blood group Γ in a hypothetical population

	Ruritanians	Others	Total
Blood group Γ	700	100	800
Others	300	18 900	19 200
Total	1000	19 000	20 000

which may be recorded as relevant background information I. (Note here that if it is thought unreasonable to be able to eliminate all innocent sources of the stain then consider, analogously, an example of a rape case in which a semen stain replaces the blood stain.) There will normally be other information which a jury, for example, will have to consider. However, in this context, I is restricted to information considered relevant in the sense that the evidence for which probabilities are of interest is dependent on I. A suspect has been identified. He is also of type Γ. Blood of type Γ is not common amongst the general population to which the criminal is thought to belong, being found in only 4% of this population. However, the suspect is discovered to be of Ruritanian ethnicity and blood of type Γ is found in 70% of Ruritanians. How, if at all, should knowledge of the suspect's ethnicity be taken into account? There is assumed at present to be no other evidence, such as eyewitness evidence, to provide information about the ethnic group of the criminal.

Consider Table 5.2. Notice that 800/20 000 (4%) of the general population are of blood group Γ whereas 700/1000 (70%) of Ruritanians are of blood group Γ, satisfying the description above.

The relevance of the Ruritanian ethnicity of the suspect can be determined by evaluating the likelihood ratio. The evidence E may be partitioned into three parts:

- E_r: the racial grouping (Ruritanian) of the suspect;
- E_s: the blood group (Γ) of the suspect;
- E_c: the blood group (Γ) of the crime stain.

The likelihood ratio is then

$$V = \frac{Pr(E \mid C)}{Pr(E \mid \overline{C})}$$

$$= \frac{Pr(E_r, E_s, E_c \mid C)}{Pr(E_r, E_s, E_c \mid \overline{C})}$$

where the two hypotheses to be compared are:

- C: there was contact between the suspect and the crime scene;
- $\bar{C}$: there was not contact between the suspect and the crime scene.

Note that explicit mention is not made of the background information I that there was no innocent explanation for the crime stain. However, it should be remembered that all the probabilities under discussion here are conditional on I. It is also assumed implicitly that if contact (C) has taken place then the evidence (blood stain) was left by the suspect. This may be thought to imply guilt (G) but the inference of guilt from contact is not one which the forensic scientist should make. Rather it is for the jury to make this inference, bearing in mind all other evidence presented at the trial.

Scene-anchored perspective

The scene-anchored perspective is one in which the suspect evidence (E_r, E_s) is conditioned on the scene evidence E_c. Using the odds form of Bayes' Theorem (2.4),

$$V = \frac{Pr(E_r, E_s, E_c \mid C)}{Pr(E_r, E_s, E_c \mid \bar{C})}$$

$$= \frac{Pr(E_r, E_s \mid C, E_c)Pr(E_c \mid C)}{Pr(E_r, E_s \mid \bar{C}, E_c)Pr(E_c \mid \bar{C})}.$$

Consider, first, the ratio $Pr(E_c \mid C)/Pr(E_c \mid \bar{C})$. If no more is assumed of the suspect other than that he was at the crime scene (C, the numerator) or that he was not at the crime scene ($\bar{C}$, the denominator) then the frequency of Γ is the same whether the suspect was present or not. Thus, $Pr(E_c \mid C) = Pr(E_c \mid \bar{C})$ and

$$V = \frac{Pr(E_r, E_s \mid C, E_c)}{Pr(E_r, E_s \mid \bar{C}, E_c)}.$$

This may be written as

$$\frac{Pr(E_r \mid C, E_c)Pr(E_s \mid E_r, C, E_c)}{Pr(E_r \mid \bar{C}, E_c)Pr(E_s \mid E_r, \bar{C}, E_c)}.$$

If the suspect was present at the crime scene (C) and if the crime stain is of group Γ (E_c) then the probability he is Ruritanian is $7/8$, the proportion of Ruritanians amongst those of group Γ. Thus $Pr(E_r \mid C, E_c) = 7/8$.

If the suspect was present at the crime scene (C) and if the crime stain is of group Γ (E_c) then the probability the suspect's blood group is Γ (E_s) is 1, independent of his ethnicity (E_r). Thus $Pr(E_s \mid E_r, C, E_c) = 1$.

If the suspect was not at the scene of the crime $(\bar{C})$ the blood group (E_c) of the crime stain gives no information about his ethnicity (E_r). Thus $Pr(E_r | \bar{C}, E_c) = Pr(E_r | \bar{C}) = 1/20$, the proportion of Ruritanians in the general population.

Similarly, if the suspect was not at the scene of the crime, the blood group of the crime stain gives no information about the blood group of the suspect. Thus $Pr(E_s | E_r, \bar{C}, E_c) = Pr(E_s | E_r, \bar{C})$ and this is the proportion of Ruritanians which are of blood group Γ or, alternatively, the probability that a Ruritanian, selected at random from the population of Ruritanians, is of blood group Γ. This probability is 7/10. Then

$$V = \frac{(\frac{7}{8}) \times 1}{(\frac{1}{20}) \times (\frac{7}{10})}$$

$$= \frac{7/8}{7/200}$$

$$= \frac{200}{8}.$$

This is the reciprocal of the proportion of people of blood group Γ in the general population. The ethnicity of the suspect is not relevant. The general proof of this result is given in Evett (1984).

Suspect-anchored perspective

The suspect-anchored perspective is one in which the scene evidence (E_c) is conditioned on the suspect evidence (E_r, E_s). From the odds form of Bayes' Theorem (2.4)

$$V = \frac{Pr(E_r, E_s, E_c | C)}{Pr(E_r, E_s, E_c | \bar{C})}$$

$$= \frac{Pr(E_c | C, E_r, E_s) Pr(E_r, E_s | C)}{Pr(E_c | \bar{C}, E_r, E_s) Pr(E_r, E_s | \bar{C})}$$

Consider the ratio $Pr(E_r, E_s | C)/Pr(E_r, E_s | \bar{C})$. Assume there is no particular predisposition towards (or away from) criminality amongst Ruritanians (E_r) or those of blood group $\Gamma (E_s)$. Then, $Pr(E_r, E_s | C) = Pr(E_r, E_s | \bar{C})$ and

$$V = \frac{Pr(E_c | C, E_r, E_s)}{Pr(E_c | \bar{C}, E_r, E_s)}.$$

The numerator $Pr(E_c | C, E_r, E_s)$ of this ratio equals 1. The suspect is assumed to have been present at the scene of the crime (C), to be Ruritanian (E_r) and of blood group Γ (E_s). The background information I, assumed implicitly, is that the criminal left the blood stain. Thus, given E_s, $Pr(E_c \mid C, E_r, E_s) = 1$. (Remember it has also been assumed that if contact C did take place then it was the suspect that left the crime stain.)

If the suspect is assumed innocent ($\bar{C}$), the information that he is Ruritanian and of blood group Γ is not of relevance for determining the probability that the crime stain is of group Γ. Thus, the estimate to be used for its probability is just $800/20\,000$ ($8/200$), its proportion in the general population. Then

$$Pr(E_c \mid \bar{C}, E_r, E_s) = 8/200,$$

$$V = 200/8.$$

As before, the ethnicity of the suspect is not relevant. Also, the scene-anchored and suspect-anchored perspective provide the same result.

Suppose now that all innocent sources of the stain have not been eliminated. Consider the scene-anchored perspective. Assume there was contact between the suspect and the crime scene (C) and that the stain at the crime scene (though not necessarily the 'crime' stain, in that it may not have been left by the criminal) is of group Γ (E_c). No information is contained in this evidence about the ethnicity of the suspect. Thus

$$Pr(E_r \mid C, E_c) = 1000/20\,000 = 1/20.$$

The stain at the crime scene may not have come from the criminal. (See Section 6.5 for a more detailed discussion of this idea, known as *relevance*, Stoney, 1991 and 1994.) Thus, the probability that the blood group of the suspect is Γ, given he is Ruritanian (E_r) and that there was contact (C) , is just the relative frequency of Γ amongst Ruritanians, which is $700/1000$ ($7/10$). Thus

$$Pr(E_s \mid E_r, C, E_c) = 7/10.$$

The probabilities in the denominator have the same values as before, namely, $Pr(E_r \mid \bar{C}, E_c) = 1/20$ and $Pr(E_s \mid E_r, \bar{C}, E_c) = 7/10$. Hence, $V = 1$. In this case the evidence has no probative value since the stain may have come from somewhere else. A similar line of argument derives the same result for the suspect-anchored perspective.

Eyewitness evidence

Consider now eyewitness evidence which states that the criminal is Ruritanian. This eyewitness evidence is assumed to be completely reliable; how to account for evidence which is less than completely reliable is not discussed. Note, however, that in such a discussion there are two conditional probabilities to be considered. Let T be an event and let W_T be an eyewitness report of T. Then it is necessary to consider both $Pr(T \mid W_T)$, the probability the event happened given the eyewitness said it did and $Pr(W_T \mid T)$, the probability the eyewitness said the event happened given it did. The purpose of including eyewitness evidence here is to illustrate the effect of restricting the population of potential offenders to a particular subgroup of a more general population.

Suppose now, once more, that the crime stain has been identified as coming from the criminal. Thus, the relevant background information, I, is in two parts: the ethnicity of the criminal and the identification of the crime stain as coming from the criminal. The evidence E is now only of two parts:

- E_s: the blood group Γ of the suspect;
- E_c: the blood group Γ of the crime stain.

Evidence E_r of the ethnicity of the suspect has been subsumed into I and is now evidence of the ethnicity of the criminal.

For the scene-anchored perspective, with I assumed implicitly,

$$V = \frac{Pr(E_s \mid C, E_c)}{Pr(E_s \mid \overline{C}, E_c)}.$$

The numerator $Pr(E_s \mid C, E_c) = 1$ since, if contact is assumed and the blood group of the crime stain is Γ, the blood group of the suspect is certain to be Γ also. The denominator $Pr(E_s \mid \overline{C}, E_c) = 7/10$, the frequency of Γ amongst Ruritanians. Hence, $V = 10/7$. The eyewitness evidence is such as to ensure that the ethnicity of the suspect (Ruritanian) is relevant.

For the suspect-anchored perspective, again with I assumed implicitly,

$$V = \frac{Pr(E_c \mid C, E_s)}{Pr(E_c \mid \overline{C}, E_s)}.$$

If the suspect was in contact with the crime scene and is of blood group Γ then the crime stain is certain to be of group Γ. Thus, the numerator, $Pr(E_c \mid C, E_s) = 1$. The denominator $Pr(E_c \mid \overline{C}, E_s) = 7/10$ since I includes the information that the criminal is Ruritanian. The frequency of blood group Γ amongst Ruritanians is $7/10$.

The scene-anchored and suspect-anchored perspectives give the same result.

5.3.2 Transfer of evidence from the scene to the criminal

This situation is analogous to the situation in the previous section when the stain at the crime scene could not be identified as coming from the criminal. It is discussed in detail in Evett (1984). The complication introduced when it cannot be assumed that the transferred particle form of the evidence is associated with the crime is made more explicit when transfer is in the direction from the scene to the criminal.

A crime has been committed during which the blood of a victim has been shed. A suspect has been identified. A single blood stain of group Γ has been found on an item of the suspect's clothing. The suspect's blood group is *not* Γ. The victim's blood group is Γ. There are two possibilities:

- A_1: the blood stain came from some background source;
- A_2: the blood stain was transferred during the commission of the crime.

As before, there are two hypotheses to consider:

- C: the suspect and the victim were in contact;
- $\bar{C}$: the suspect and the victim were not in contact.

The evidence E to be considered is that a single blood stain has been found on the suspect's clothing and that it is of group Γ. The information that the victim's blood group is Γ is to be considered as part of the relevant background information I. This is a scene-anchored perspective. The value of the evidence is then

$$V = \frac{Pr(E \mid C, I)}{Pr(E \mid \bar{C}, I)}.$$

Consider the numerator first and event A_1 initially. Then, the suspect and the victim have been in contact (C) and no blood has been transferred to the suspect. This is an event with probability $Pr(A_1 \mid C)$. Also, a stain of group Γ has been transferred by some other means, an event with probability $Pr(B, \Gamma)$ where B refers to the event of a transfer of a stain from a source (the background source) other than the crime scene.

Consider A_2. Blood has been transferred to the suspect, an event with probability $Pr(A_2 \mid C)$; given A_2, C and the blood group Γ of the victim, it is certain that the group of the transferred stain is Γ. This implies also that no blood has been transferred from a background source.

5.3.3 Transfer probabilities

The probability that no blood has been transferred by other means has also to be included. Let $t_0 = Pr(A_1 \mid C)$ and $t_1 = Pr(A_2 \mid C)$ denote the probabilities of no stain or one stain being transferred during the course of a crime. Let $b_\rho(0)$ and $b_\rho(1)$, respectively, denote the probabilities that a person from the relevant population will have zero blood stains or one blood stain on his clothing. The subscript ρ is used to indicate that the probabilities are taken with respect to an object, here a person but not necessarily always so, which has received the evidence. Such a body is a *receptor* (Evett, 1984, and Section 1.4). Let $p_\rho(\Gamma)$ denote the probability that a stain acquired innocently on the clothing of a person from the relevant population will be of group Γ. This probability may be different from the relative frequency of Γ amongst the general population (Gettinby, 1984). Then $Pr(B, \Gamma) = p_\rho(\Gamma)b_\rho(1)$. The numerator can be written as (Evett, 1984)

$$t_0 p_\rho(\Gamma)b_\rho(1) + t_1 b_\rho(0).$$

The first term accounts for A_1, the second for A_2.

Now consider the denominator. The suspect and the victim were not in contact. The denominator then takes the value $Pr(B, \Gamma)$ which equals

$$p_\rho(\Gamma)b_\rho(1).$$

The value of the evidence is thus

$$V = \frac{\{t_0 p_\rho(\Gamma)b_\rho(1) + t_1 b_\rho(0)\}}{\{p_\rho(\Gamma)b_\rho(1)\}}$$

$$= t_0 + \frac{t_1 b_\rho(0)}{b_\rho(1)b_\rho(\Gamma)}. \tag{5.4}$$

Consider the probabilities which have to be estimated:

- t_0: no stain transferred during the commission of a crime;
- t_1: one stain transferred during the commission of a crime;
- $b_\rho(0)$: no stain transferred innocently;
- $b_\rho(1)$: one stain transferred innocently;
- $p_\rho(\Gamma)$: the frequency of blood of group Γ amongst blood stains on clothing.

The first four of these probabilities relate to what has been called *extrinsic evidence*, the fifth to *intrinsic evidence* (Kind, 1994). The estimation of probabilities for extrinsic evidence is subjective and values for these are matters for a forensic scientist's personal judgement. The estimation of probabilities for intrinsic evidence may be determined by observation and measurement. Note that, in general, t_0 is small in relation to $t_1 b_\rho(0)/\{b_\rho(1)p_\rho(\Gamma)\}$ and may be considered negligible.

The following example is taken from Evett (1984). Let Γ be the blood group combination *AB, Hp 2–2* which has a relative frequency of approximately 0.01 in England (Stedman, 1972). Assume that the distribution of blood groups among stains on clothing is approximately the distribution among the relevant population. This assumption is not necessarily correct (Gettinby, 1984) and there further discussion of it below. Then $p_\rho(\Gamma) = 0.01$. A survey of men's clothing was conducted by Briggs (1978) from which it appears reasonable that $b_\rho(0) > 0.95$, $b_\rho(1) < 0.05$. The transfer probabilities t_0 and t_1 require to be estimated from a study of the circumstances of the crime and, possibly, experimentation. Suppose $t_1 > 0.5$ (Evett, 1984). Then, irrespective of the value of t_0 (except that it has to be less than $(1 - t_0)$; i.e., less than 0.5 in this instance),

$$V > \frac{0.5 \times 0.95}{0.05 \times 0.01} = 950,$$

a value which indicates very strong evidence to support the hypothesis that the suspect and the victim were in contact. The evidence is at least 950 times more likely if the suspect and victim were in contact than if they were not.

Notice that (5.4) is considerably different from $1/p_\rho(\Gamma)$ ($= 100$ in the numerical example). This latter result would hold if $(t_1 b_0/b_1)$ were approximately 1, which may mean that unrealistic assumptions about the relative values of the transfer probabilities would have to be made.

Bloodstains on clothing

The assumption that the distribution of blood groups among stains on clothing is approximately the distribution among the relevant population has been questioned (Gettinby, 1984). This is because the blood on a piece of clothing may have come from the wearer of the clothing and thus there is a bias in favour of the blood group of the wearer. The following argument is based on that in Gettinby (1984).

Consider a population of size N in which a proportion p have innocently acquired bloodstains on their clothing. Let p_o, p_a, p_b and p_{ab} be the proportions with which blood groups O, A, B and AB occur in the population such that $p_o + p_a + p_b + p_{ab} = 1$.

Consider people of group O. Bloodstains detected on clothing may arise from several sources:

- by self-transfer (O type stains), with probability α, say,
- O stains from somewhere else, with probability β_0, say,
- stains of type A, B or AB, necessarily from somewhere else, with probability γ_0, say,

such that

$$\alpha + \beta_0 + \gamma_0 = 1,$$
$$\beta_0 = (1 - \alpha)p_o. \qquad (5.5)$$

The proportion α is independent of the blood grouping of the individuals under consideration, unlike β_0 and γ_0.

With an intuitively obvious notation, the following results for individuals of types A, B and AB can be stated:

$$\alpha + \beta_a + \gamma_a = 1,$$
$$\beta_a = (1 - \alpha)p_a; \tag{5.6}$$

$$\alpha + \beta_a + \gamma_b = 1,$$
$$\beta_b = (1 - \alpha)p_b; \tag{5.7}$$
$$\alpha + \beta_{ab} + \gamma_{ab} = 1,$$
$$\beta_{ab} = (1 - \alpha)p_{ab}. \tag{5.8}$$

Of those individuals who have bloodstains which have arisen from a source other than themselves (*non-self stains*) only a proportion γ will be distinguishable as such, where

$$\gamma = p_o\gamma_o + p_a\gamma_a + p_b\gamma_b + p_{ab}\gamma_{ab}.$$

For example $p_o\gamma_o = Pr(\text{type } A, B \text{ or } AB \text{ stain found on clothing} \mid \text{person is of type } O) \times Pr(\text{person is of type } O)$. Multiplication of pairs of equations (5.5) to (5.8) by p_o, p_a, p_b and p_{ab}, respectively, gives:

$$p_o\alpha + (1 - \alpha)p_o^2 + p_o\gamma_o = p_o$$
$$p_a\alpha + (1 - \alpha)p_a^2 + p_a\gamma_a = p_a$$
$$p_b\alpha + (1 - \alpha)p_b^2 + p_b\gamma_b = p_b$$
$$p_{ab}\alpha + (1 - \alpha)p_{ab}^2 + p_{ab}\gamma_{ab} = p_{ab}$$

and summing gives

$$\alpha + (1 - \alpha)(1 - \delta) + \gamma = 1 \tag{5.9}$$

where

$$\delta = 1 - p_o^2 - p_a^2 - p_b^2 - p_{ab}^2$$

the discriminating power (Section 4.4) of the *ABO* system. From (5.9),

$$\alpha = 1 - \frac{\gamma}{\delta}.$$

Values of $\gamma = 0.182$ and $\delta = 0.602$ are used by Gettinby (1984) who cites Briggs (1978). From these a value of $\alpha \simeq 0.7$ is obtained for the estimate of the probability of a bloodstain being acquired from oneself, given that a bloodstain has been found on the clothing; i.e., approximately 70% of bloodstains on clothing are acquired by self-transfer.

Consider a person of blood group O. Denote the probability that he innocently bears a bloodstain and the bloodstain is of type O by $C_0(O)$. Then

$$C_0(O) = Pr(\text{suspect has stain from self})$$

$$+ Pr(\text{suspect has stain not from self but of type } O)$$

$$= p\alpha + p(1-\alpha)p_o.$$

With a similar notation, for a person of blood group O to bear bloodstains of type A, B or AB, the probabilities are

$$C_A(O) = p(1-\alpha)p_a,$$

$$C_B(O) = p(1-\alpha)p_b,$$

$$C_{AB}(O) = p(1-\alpha)p_{ab}.$$

The sum

$$C_0(O) + C_A(O) + C_B(O) + C_{AB}(O) = p,$$

the probability of innocently acquiring a bloodstain. A value of $p = 0.369$ is given by Briggs (1978) and used by Gettinby (1984). Also, the distribution of bloodgroups amongst innocently acquired bloodstains on clothing of people of type O may be determined. For example, the probability a person of type O has a bloodstain of type O on his clothing, given it was acquired innocently is $C_0(O)/p = \alpha + (1-\alpha)p_o$. The distribution of bloodgroups, for people of type O is thus:

$$Pr(\text{type } O \,|\, \text{innocently acquiring a bloodstain}) = \alpha + (1-\alpha)p_o,$$

$$Pr(\text{type } A \,|\, \text{innocently acquiring a bloodstain}) = (1-\alpha)p_a,$$

$$Pr(\text{type } B \,|\, \text{innocently acquiring a bloodstain}) = (1-\alpha)p_b,$$

$$Pr(\text{type } AB \,|\, \text{innocently acquiring a bloodstain}) = (1-\alpha)p_{ab},$$

with similar results for people of type A, B and AB. The comparison of this distribution with the general distribution is made in Table 5.3.

Populations

The importance of the population from which transfer evidence may have come has been long recognised. In 1935, it was remarked that

Table 5.3 Distribution of bloodgroups of innocently acquired bloodstains on clothing of people of type O, compared with the distribution in the general population

Bloodgroup	O	A	B	AB	Total
Clothing of people of type O	$\alpha + (1-\alpha)p_o$	$(1-\alpha)p_a$	$(1-\alpha)p_b$	$(1-\alpha)p_{ab}$	1
General population	p_o	p_a	p_b	p_{ab}	1

One need only consider the frequency with which evidence regarding blood and semen stains is produced in court to realise the need for data relating to the relative frequency of occurrence of such stains on garments in no wise related to crimes; for example, on one hundred garments chosen at random from miscellaneous sources, how many would show blood stains, how many semen stains? Questions such as these must arise in court, and answers based on experimental investigation would prove of considerable value in assessing evidence of this type.'
(Tryhorn, 1935, quoted by Owen and Smalldon, 1975.)

Various data have been published regarding what has been called *environment specific population data* concerned with the incidence of the transferred particle form of materials on clothing (Stoney, 1991). These include glass and paint fragments on clothing (Pearson *et al.*, 1971; Dabbs and Pearson, 1970, 1972; Pounds and Smalldon, 1978; Harrison *et al.*, 1985; McQuillan and Edgar, 1992), glass in footwear (Davis and DeHaan, 1977; Walsh and Buckleton, 1986), loose fibres on clothing (Owen and Smalldon, 1975; Briggs, 1978; Fong and Inami, 1986). These surveys are relevant when the material associated with the suspect is in particle form on clothing. If the material is in source form, then population data for the source form are needed.

An example of the importance of the choice of the correct population is given by Walsh and Buckleton (1986), already discussed in Section 5.2. A crime is committed in which a window is broken. The refractive index of glass from the scene of the crime is 1.5202. A suspect is identified and he has fragments of glass embedded in his footwear whose refractive index is also 1.5202. However, the glass in the footwear is container glass of which 3.1% has a refractive index of 1.5202 whilst only 0.23% of building glass in the window has a refractive index of 1.5202, these percentages coming from Lambert and Evett (1984) reporting on U.K. glass. The frequency which is relevant for assessing the value of the evidence E is 3.1%, giving a probability of 0.031 of finding E on an innocent ($\bar{G}$) person, $Pr(E \mid \bar{G})$. Notice that the relevant population is that which is applicable to innocent people. This is so because it is the population from which the transferred particle may have come.

Consider, however, transfer to the scene of a blood stain. The relevant population is that from which the criminal has come, and has nothing to do with any population which may be defined by the suspect.

The choice of population is not so clear, however, in surveys conducted into bloodstains on clothing. The survey by Owen and Smalldon (1975) was conducted by visiting a dry cleaning establishment. Briggs (1978) used data from two large-scale murder investigations during the course of which large numbers of articles of clothing from numerous male suspects were examined in the laboratory for the presence of bloodstaining. In these investigations, Briggs (1978) argued, very reasonably, that the sampling was biased in that it was all derived from people involved in a crime investigation. This may be expected to lead to a higher incidence of groupable bloodstains on clothing than a survey of clothing from the general population would give. Briggs (1978) reported that, for the *ABO* blood group, in the second murder investigation, a total of 966 items of clothing from 122 suspects were examined. Forty five of these had blood on their clothing $(45/122 = 0.369$, the value for p used by Gettinby, 1984). There were 22 suspects for which a grouping result was obtained from their clothing and of these only four provided clothing with stains where the grouping result differed from the donor's blood group; this gives a value of $4/22$ (0.182) for γ as used by Gettinby (1984). From the original 122 suspects, for only four (3.2%), with nine articles, was it possible to show the presence of blood which did not originate from the wearer. Eight of these nine articles originated from three of the four suspects, all three of whom had a history of violence. The conclusion drawn by Briggs was that the proportion of people who have blood on their clothing, of a different group from their own, is 3.2%. Gettinby's parameter γ is the proportion of those people who have bloodstains on their clothing which is not of their own blood where the bloodstain is distinguishable as such: if n is the number of people who have bloodstains on their clothing which is not of their own blood and n_0 is the number of people who have bloodstains on their clothing which is not of their own blood and is of a different blood group (and so is distinguishable as coming from someone else), then $\gamma = n_o/n$.

Owen and Smalldon (1975) reported that out of 100 pairs of trousers examined at the dry cleaning establishment 16 (16%) had bloodstains on them. However, it was not possible to determine if the blood group was different from that of the wearer. For 100 jackets, 5% had bloodstains on them. For trousers, 44% had semen stains.

5.3.4 Presence of non-matching evidence

In transfer evidence it may be that there is evidence present on the suspect or at the scene of the crime which does not match that found at the scene of the crime or on the suspect, respectively. For example, consider a case involving the transfer of fibres from the scene of a crime to the criminal. A suspect is found with fibres on his clothing which match, in some sense, fibres from the scene of the crime. However, he also has fibres of many different types on his clothing which do not match those found at the scene (so-called *foreign* fibres).

A likelihood ratio for such a situation has been derived by Grieve and Dunlop (1992). It includes factors to allow for transfer probabilities, probabilities of foreign fibre types being found on the person, relative frequencies of occurrence of the matching fibre types and a factor to account for the number of matching fibres amongst the total number of fibres. There is considerable subjectivity in the determination of these figures. The importance of the work lies in the recognition that the number of items found which do *not* match has to be considered when assessing the evidence as well as the number of items found which do match.

5.4   RELEVANT POPULATIONS

A similarity is found between transfer evidence (E_c) of a trait (e.g., blood grouping) found at the crime scene and evidence of the same trait found on the suspect (E_s). The value of the similarity between E_c and E_s is assessed partly by comparison with respect to some population. The similarity may be purely coincidental. A more general derivation of the results in Sections 5.3.1 and 5.3.2 is now given.

Transfer from the criminal to the scene of the crime

A suspect-anchored perspective is taken. The value of the evidence is given by

$$V = \frac{Pr(E_c, E_s \mid C)}{Pr(E_c, E_s \mid \bar{C})}$$

$$= \frac{Pr(E_c \mid E_s, C)}{Pr(E_c \mid E_s, \bar{C})} \times \frac{Pr(E_s \mid C)}{Pr(E_s \mid \bar{C})}.$$

When there is transfer from the criminal to the scene, the evidence found on the suspect, E_s, is independent of whether the suspect was present (C) at the scene of the crime or not ($\bar{C}$). Thus,

$$Pr(E_s \mid C) = Pr(E_s \mid \bar{C})$$

and

$$V = \frac{Pr(E_c \mid E_s, C)}{Pr(E_c \mid E_s, \bar{C})}.$$

If $\bar{C}$ is true then E_c is independent of E_s and thus

$$Pr(E_c \mid E_s, \bar{C}) = Pr(E_c \mid \bar{C}).$$

Assume for the present that $Pr(E_c \mid E_s, C) = 1$. This is a reasonable assumption for bloodgroup frequency data, for example. Then

$$V = \frac{1}{Pr(E_c \mid \bar{C})}.$$

Transfer from the scene of the crime to the criminal

A scene-anchored perspective is taken. The value of the evidence is given by

$$V = \frac{Pr(E_c, E_s \mid C)}{Pr(E_c, E_s \mid \bar{C})}$$

$$= \frac{Pr(E_s \mid E_c, C)}{Pr(E_s \mid E_c, \bar{C})} \times \frac{Pr(E_c \mid C)}{Pr(E_c \mid \bar{C})}.$$

The evidence at the scene of the crime, E_c, is independent of whether the suspect was present (C) at the scene or not ($\bar{C}$). Thus,

$$Pr(E_c \mid C) = Pr(E_c \mid \bar{C})$$

and

$$V = \frac{Pr(E_s \mid E_c, C)}{Pr(E_s \mid E_c, \bar{C})}.$$

If $\bar{C}$ is true, then E_s is independent of E_c and

$$Pr(E_s \mid E_c, \bar{C}) = Pr(E_s \mid \bar{C}).$$

Assume for the present that $Pr(E_s \mid E_c, C) = 1$. Then

$$V = \frac{1}{Pr(E_s \mid \bar{C})}.$$

The results from the two perspectives may be represented in one result by noticing that in both cases the evidence of interest is the transferred particle form. Thus, if E_{tp} denotes transferred particle form of the evidence then

$$V = \frac{1}{Pr(E_{tp} \mid \bar{C})} \tag{5.10}$$

With reference to which population is $Pr(E_{tp} \mid \bar{C})$ to be evaluated? In the case of blood group evidence found at the scene of the crime, the population could be based on the ethnic group of the suspect but this is incorrect as earlier discussion (Sections 5.3.1 and 5.3.3) has shown. The concept of a *relevant population* was introduced by Coleman and Walls (1974) who said that 'the relevant population are those persons who could have been involved; sometimes it can be established that the crime must have been committed by a particular class of persons on the basis of age, sex, occupation or other sub-grouping, and it is then not necessary to consider the remainder of, say, the United Kingdom'. The concept of a *suspect population*, defined as 'the smallest population known to possess the culprit as a member' was introduced by Smith and Charrow (1975). The term *suspect population* was also used by Lempert (1991) who used it to refer to the population of possible perpetrators of the crime. These populations should not be confused with any population which may be defined by reference to the suspect only, despite their name. It could be argued that the smallest 'suspect population' is just the population of the world but this is not helpful.

The *total population involved* as an ideal of a population suitable for study in order to evaluate significance was suggested by Kirk and Kingston (1964). An *appropriate population* from which a sample can be taken for the study was suggested by Kingston (1965a) without further elaboration of the meaning of 'appropriate'. The estimation of the expected number of people in a population which would be expected to have a characteristic in question was considered by Kingston (1965b) who concluded it was best to base calculations on the *maximum possible population*.

Evaluation of fingerprint evidence is not discussed here. It is a very complex issue, beyond the scope of this book, and there have been only a few attempts to construct a suitable statistical model (Osterburg *et al.*, 1977; Sclove, 1979, 1980; Stoney and Thornton, 1986; Hrechak and McHugh, 1990; Mardia *et al.*, 1992). However, some of the debate concerning the definition of a suitable population for the assessment of fingerprint evidence is of interest here. Suppose a suspect has been chosen on the basis of fingerprint evidence alone. The underlying population from which the suspect has to be considered as being selected has been argued as being that of the whole world (Kingston, 1964). However, Stoney and Thornton (1986) argued that it is rarely the case that a suspect would be chosen purely on the basis of fingerprint evidence. Normally, there would have been a small group of suspects which would have been isolated from the world population on the basis of other evidence, though Kingston (1988) disagreed with this. The fingerprint evidence has then to be considered relative to this small group only. Further discussion of this point is given in Aitken (1991b).

More recent discussion about populations in the context of the island problem (Section 2.5.5, Eggleston, 1983; Yellin, 1979; Lindley, 1987) including the effect of search procedures on estimations of the probability of guilt through reduction of the population size may be found in Balding and Donnelly (1994) and Dawid (1994).

Notice that a population need not necessarily be described by a geographical boundary. The distribution of genetic markers in United States populations has

Table 5.4 Probability that a criminal lived in a particular area, given the crime was committed in Auckland, New Zealand

Residence of offender	Auckland	Remainder of North Island	Remainder of New Zealand	Remainder of the world
Approximate population	809 000	1603 000	1660 000	$\simeq 4000\,000\,000$
Probability	0.92	0.05	0.03	0.00

been studied by Gaensslen *et al*, (1987a, b, c) where populations were defined by 'the effective interbreeding gene pool'.

A suitable name for the population in question is required. One possible though rather lengthy name is the *potential perpetrator population*. When there is information about the source of the transfer evidence then $Pr(E_{tp} | \bar{C})$ can be calculated with respect to a *potential source population*. In many cases these two populations are approximately the same. However, Koehler (1993) cites the novel '*Presumed Innocent*' by Scott Turow in which a woman commits a murder and plants her husband's semen in the victim in an effort to incriminate him. The woman may be a member of the potential perpetrator population but not of the potential source population. Alternatively, consider a case, in which a woman is murdered in her bed one week after her husband died (Koehler, 1993). Hairs recovered at the scene of the crime may belong to the woman's deceased husband. This would place him in the potential source population but he would not be a member of the potential perpetrator population. Further details with reference to blood grouping are given in Chapter 6.

As will be seen in Chapter 8 on DNA profiling, for a case involving the analysis of DNA evidence, the consideration of the close relatives of a suspect as part of the potential source population will lead to values of $Pr(E_{tp} | \bar{C})$ larger than values based on a general population, and hence to smaller values for V. A suspect's relatives are more likely to be genetically similar to the suspect than are random members of a general population. The difficulty of defining a potential source population may cause problems with the evaluation of $Pr(E_{tp} | \bar{C})$. If close relatives are included then there could be dramatic alterations in $Pr(E_{tp} | \bar{C})$.

An interesting contribution to the discussion is made by Walsh and Buckleton (1994). They report the results of a study which attempts to estimate the general place of residence of a criminal, given a crime was committed in Auckland, New Zealand. These are given in Table 5.4.

In the absence of any information, including that of its location, about the crime the prior odds in favour of contact between the suspect and the crime scene are $Pr(C)/Pr(\bar{C}) = 1/N$ where N is the population of the world, as argued by Kingston (1964). Consider information I which is in two parts:

- I_1: the offence was committed in Auckland;
- I_2: the suspect lives in Auckland.

Thus I, the other information, reference to which has usually been supressed, may now be written as $I = (I_1, I_2)$ and the prior odds in favour of C, given I, may be written as

$$\frac{Pr(C \mid I_1, I_2)}{Pr(\overline{C} \mid I_1, I_2)}.$$

Then, it can be shown (Walsh and Buckleton, 1994) that

$$\frac{Pr(C \mid I_1, I_2)}{Pr(\overline{C} \mid I_1, I_2)} \times \frac{Pr(I_1, I_2)}{Pr(I_1, I_2)} = \frac{Pr(I_1, I_2, C)}{Pr(I_1, I_2, \overline{C})}$$

$$= \frac{Pr(I_1, I_2 \mid C)}{Pr(I_1, I_2 \mid \overline{C})} \times \frac{Pr(C)}{Pr(\overline{C})}$$

$$= \frac{Pr(I_2 \mid I_1, C) Pr(I_1 \mid C)}{Pr(I_1 \mid I_2, \overline{C}) Pr(I_2 \mid \overline{C})} \times \frac{Pr(C)}{Pr(\overline{C})}.$$

Assume that $Pr(I_1 \mid C) = Pr(I_1 \mid I_2, \overline{C}) = Pr(I_1)$. In other words, it is assumed, firstly, that the probability that the crime was committed in Auckland is independent of whether the suspect committed the crime or not (since nothing else is assumed about the suspect, in particular it is not assumed that the suspect lives in Auckland). Similarly, it is assumed that the probability that the crime was committed in Auckland is independent of the information that the suspect lived in Auckland given that the suspect was not in contact with the crime scene.

It is also assumed that $Pr(I_2 \mid \overline{C}) = Pr(I_2)$; i.e., the probability that the suspect lives in Auckland given he was not in contact with the crime scene is equal to the probability that he lives in Auckland. Hence

$$\frac{Pr(C \mid I_1, I_2)}{Pr(\overline{C} \mid I_1, I_2)} = \frac{Pr(I_2 \mid I_1, C)}{Pr(I_2)} \times \frac{Pr(C)}{Pr(\overline{C})}.$$

With the results from Table 5.4, $Pr(I_2 \mid I_1, C) = 0.92$ and $Pr(I_2) \simeq 809\,000/4\,000\,000\,000$. Also, $Pr(C)/Pr(\overline{C}) = 1/4000\,000\,000$. Hence

$$\frac{Pr(C \mid I_1, I_2)}{Pr(\overline{C} \mid I_1, I_2)} = \frac{0.92}{809\,000} = 1.14 \times 10^{-6}.$$

The prior odds of 1/(population of the world) have been considerably reduced. However, the odds of 1/(population of Auckland) have also been altered by a factor of 0.92 to account for the information that the suspect lived in Auckland. This seems intuitively reasonable.

This discussion illustrates the considerable problems surrounding the definitions of populations and the choice of suitable names. For example, the distinction between a potential perpetrator population and a potential source population needs to be borne in mind.

In general discussion, the term *relevant population* will be used and this will have the following rather technical definition

Definition   The *relevant population* is the population defined by the combination of the hypothesis $\bar{C}$ proposed by the defence and the background information I.

For transfer from the scene to the criminal, I may include information about the suspect. For example, with blood stain evidence found on the suspect's clothing, the lifestyle of the suspect is relevant as is his ethnic group (Gettinby, 1984, and Section 5.3.3). For transfer from the criminal to the scene, $\bar{C}$ dissociates the suspect from the scene. The population from which the evidence may have come and which should be used for the determination of V is specified with the help of I.

5.5   ERRORS IN GROUPING

Consider a case in which a blood stain on the clothing of a suspect is found to match that of the victim in the sense that it is of the same group over each of several genetic systems. Let there be m such systems, with k_i phenotypes in the ith system $(i = 1, \ldots, m)$. Let the match be in groups which occur with *true* probability p_{ij_i}, $(j_i = 1, \ldots, k_i)$ in the ith system, independently of other systems. The probability θ of a match over each of the systems is therefore

$$\theta = p_{1j_1} p_{2j_2} \cdots p_{mj_m} = \prod_{i=1}^{m} p_{ij_i}.$$

This probability is sometimes known as the *probability of concordance* (Selvin and Grunbaum, 1987). It is correct if the typing were error-free.

Suppose, now, that there are errors in the typing. A laboratory misclassifies phenotypes within the ith system such that Pr(subject is classified as phenotype other than j_i|subject is of phenotype j_i) $= e_i$. Thus Pr(subject is classified as phenotype j_i | subject is of phenotype j_i) $= (1 - e_i)$. Let $p^*_{ij_i}$ be the *apparent* probability of classifying an individual as having phenotype j_i in the ith system. Assume that a misclassification of one of the other $(k_i - 1)$ phenotypes as phenotype j_i is with equal probability $1/(k_i - 1)$; thus, the probability a subject is *classified* as phenotype j_i, conditional on the subject's true phenotype being other than j_i, is $e_i/(k_i - 1)$. Then, let:

- s_{j_i} denote subject is of phenotype j_i;
- $s^*_{j_i}$ denote subject is classified as j_i;
- $\bar{s}_{j_i}$ denote subject is not of phenotype j_i.

Thus, from above, $Pr(s^*_{j_i} | \bar{s}_{j_i}) = e_i/(k_i - 1)$.

$$
\begin{aligned}
p^*_{ij_i} &= Pr(s^*_{j_i}) \\
&= Pr(s^*_{j_i} | s_{j_i})Pr(s_{j_i}) + Pr(s^*_{j_i} | \bar{s}_{j_i})Pr(\bar{s}_{j_i}) \\
&= (1 - e_i)p_{ij_i} + e_i(1 - p_{ij_i})/(k_i - 1).
\end{aligned}
$$

Consider the following two hypotheses:

- C; the stain on the suspect's clothing came from the victim;
- $\bar{C}$: the stain on the suspect's clothing did not come from the victim.

Note that this is a simplification of the example above from Evett (1984) in which the hypotheses were of contact and of no contact. Evett included transfer probabilities in his solution. It is straightforward to include the result for an error-prone laboratory into Evett's solution.

For an error-free laboratory, the value of the evidence is

$$
V = \frac{1}{\theta}
$$

Consider an error-prone laboratory. When there are m independent systems, the probability of no error $= (1 - e_1) \cdots (1 - e_m) = \prod_{i=1}^{m}(1 - e_i)$. The overall probability of error, e, is defined to be the probability of at least one error which is the complement of the probability of no error. Thus

$$
e = 1 - \prod_{i=1}^{m}(1 - e_i) \simeq \sum_{i=1}^{m} e_i.
$$

The numerator then takes the value $(1 - e)$. Let $\theta^* = \prod_{i=1}^{m} p^*_{ij_i}$ be the probability, in an error-prone laboratory, that a blood stain on a suspect's clothing is found to match the victim. Then the value of the evidence in such a laboratory is given by

$$
V^* = \frac{(1 - e)}{\theta^*}.
$$

It can be shown that $V^* < V$ (Morris and Brenner, 1988) as follows:

$$
\begin{aligned}
p^*_{ij_i} &= (1 - e_i)p_{ij_i} + e_i(1 - p_{ij_i})/(k_i - 1) \\
&> (1 - e_i)p_{ij_i},
\end{aligned}
$$

Table 5.5 Phenotypic frequencies and error probabilities for eight blood group systems

System	Phenotype	Frequency(p_{ij_i})	Error (e_i)	k_i	$p^*_{ij_i}$
ABO	OO	0.441	0.063	4	0.425
Phosphoglucomutase	1–1	0.591	0.028	3	0.580
Esterase *D*	1–1	0.781	0.032	3	0.760
Erythrocyte acid phosphatase	*AB*	0.400	0.021	6	0.394
Adenosine deaminase	1–1	0.906	0.003	3	0.903
Adenylate kinase	1–1	0.925	0.007	3	0.919
Glyoxalase *I*	1–2	0.492	0.052	3	0.480
Haptoglobin	1–2	0.487	0.034	3	0.479

and thus

$$\frac{1}{p_{ij_i}} > \frac{(1 - e_i)}{p^1_{j_i}}.$$

For *m* independent systems

$$V = \frac{1}{\theta}$$

$$= \frac{1}{\prod_{i=1}^{m} p_{ij_i}}$$

$$> \frac{\prod_{i=1}^{m}(1 - e_i)}{\prod_{i=1}^{m} p^*_{ij_i}}$$

$$= \frac{1 - e}{\theta^*}$$

$$= V^*.$$

A result in which a match is found with several independent systems provides evidence of less value in an error-prone laboratory than in an error-free laboratory.

The following example (see Table 5.5) uses values from Selvin and Grunbaum (1987) where the error probabilities $\{e_i\}$ are given from Sensabaugh (1984):

$$\theta = 0.441 \times 0.591 \times \cdots \times 0.487$$

$$= 0.0163;$$

$$V = 61.3;$$

$$\theta^* = 0.425 \times 0.580 \times \cdots \times 0.479$$

$$= 0.0141.$$

The overall error probability $e = 1 - (0.937 \times 0.972 \times \cdots \times 0.966) = 1 - 0.783 = 0.217$ and thus

$$V^* = \frac{0.783}{0.0141} = 55.5.$$

The result that the value of the evidence is greater for an error-free laboratory than for an error-prone laboratory has been verified numerically.

These results can be incorporated into (5.4) by replacing $p_\rho(\Gamma)$ by θ for an error-free laboratory and by $\theta^*/(1-e)$ for an error-prone laboratory.

6

Discrete Data

6.1 NOTATION

As usual, let E be the evidence, the value of which has to be assessed. Let the two hypotheses to be compared be denoted C and $\bar{C}$. The hypotheses will be stated explicitly for any particular context. The likelihood ratio, V, is then

$$V = \frac{Pr(E|C)}{Pr(E|\bar{C})}.$$

However, background information I has also to be considered. The likelihood ratio is then

$$V = \frac{Pr(E|C, I)}{Pr(E|\bar{C}, I)},\tag{6.1}$$

(see (2.10), Section 2.5.1).

A discussion of the interpretation of the evidence of the transfer of a single blood stain from the criminal to the scene of the crime, following that of Section 5.3.1, is given. This is then extended to cases involving several blood stains and several offenders.

6.2 SINGLE SAMPLE

6.2.1 Introduction

From Section 2.5.1, consider

$$\frac{Pr(C|E, I)}{Pr(\bar{C}|E, I)} = \frac{Pr(E|C, I)}{Pr(E|\bar{C}, I)} \times \frac{Pr(C|I)}{Pr(\bar{C}|I)}.$$

Consider Example 1.1 in which a blood stain has been left at the scene of the crime by the person who committed the crime. A suspect has been identified and it is desired to establish the strength of the link between the suspect and the crime. A comparison between the blood group of the stain and the blood group of a sample given by the suspect is made by a forensic scientist. The two hypotheses to be compared are:

- C: there was contact between the suspect and the crime scene;
- $\bar{C}$: there was not contact between the suspect and the crime scene.

The scientist's results, denoted by E may be divided into two parts (E_c, E_s) as follows:

- E_s: the blood group, Γ, of the suspect;
- E_c: the blood group, Γ, of the crime stain.

A particular example was discussed in Section 5.3.1. A general formulation of the problem is given here. The scientist knows, in addition, from data previously collected that blood of group Γ occurs in $100\gamma\%$ of some population, Ψ say.

The value to be attached to E is given by

$$V = \frac{Pr(E|C, I)}{Pr(E|\bar{C}, I)}.$$

This can be simplified:

$$V = \frac{Pr(E|C, I)}{Pr(E|\bar{C}, I)}$$

$$= \frac{Pr(E_c, E_s|C, I)}{Pr(E_c, E_s|\bar{C}, I)}$$

$$= \frac{Pr(E_c|E_s, C, I)\,Pr(E_s|C, I)}{Pr(E_c|E_s, \bar{C}, I)\,Pr(E_s|\bar{C}, I)}.$$

Now, E_s is the evidence that the suspect's blood group is Γ. As in Section 5.3.1 it is assumed that a person's blood group is independent of whether he was at the scene of the crime (C) or not $(\bar{C})$. Thus

$$Pr(E_s|C, I) = Pr(E_s|\bar{C}, I)$$

and so

$$V = \frac{Pr(E_c|E_s, C, I)}{Pr(E_c|E_s, \bar{C}, I)}.$$

If the suspect was not at the scene of the crime ($\bar{C}$ is true) then the evidence (E_c)

about the blood group of the crime scene is independent of the evidence (E_s) about the blood group of the suspect. Thus

$$Pr(E_c|E_s, \bar{C}, I) = Pr(E_c|\bar{C}, I)$$

and

$$V = \frac{Pr(E_c|E_s, C, I)}{Pr(E_c|\bar{C}, I)}.$$

Notice that the above argument takes a suspect-anchored perspective. It is possible to consider a scene-anchored perspective. A similar argument to that used above shows that

$$V = \frac{Pr(E_s|E_c, C, I)}{Pr(E_s|\bar{C}, I)}.$$

This result assumes that $Pr(E_c|C, I) = Pr(E_c|\bar{C}, I)$, i.e., the blood group of the crime stain is independent of whether the suspect was present at the crime scene or not (remembering that nothing else is known about the suspect, in particular his blood group). The assumption that the suspect's characteristics are independent of whether he committed the crime or not should not be made lightly. It is correct with reference to the suspect's blood group. However, some crime scenes may be likely to transfer materials to an offender's clothing, for example. If the characteristics of interest relate to such materials and the offender is later identified as a suspect, the presence of such material is not independent of his presence at the crime scene. If the legitimacy of the simplifications are in doubt then the original expression (6.1) is the one which should be used.

The background information, I, may be used to assist in the determination of the relevant population from which the criminal may be supposed to have come. For example, consider an example from New Zealand where I may include an eyewitness description of the criminal as Chinese. This is valuable because the frequency of blood groups can vary between ethnic groups and affect the value of the evidence; see Section 5.3.1.

First, consider the numerator $Pr(E_c|C, E_s, I)$ of the likelihood ratio in the suspect-anchored perspective. This is the probability the crime stain is of group Γ given the suspect was present at the crime scene *and* the suspect is of blood group Γ *and* all other information, including, for example, an eyewitness account that the criminal is Chinese. This probability is just 1 since if the suspect was at the crime scene and is of blood group Γ then the crime stain is of group Γ, assuming as before that all innocent sources of the crime stain have been eliminated. Thus $Pr(E_c|C, E_s, I) = 1$.

Now consider the denominator $Pr(E_c|\bar{C}, I)$. Here the hypothesis $\bar{C}$ is assumed to be true; i.e., the suspect was not present at the crime scene. I is also assumed known. Together I and $\bar{C}$ define the relevant population (see Section 5.4).

6.2.2   General population

Suppose, initially, I provides no information about the criminal which will affect the probability of his blood group being of a particular type. For example, I may include eyewitness evidence that the criminal was a tall, young male. However, blood group is independent of all three of these qualities, so I gives no information affecting the probability of the blood group being of a particular type.

It is assumed the suspect was not at the crime scene. Thus, the suspect is not the criminal. The relevant population (Section 5.4) is deemed to be Ψ. The criminal is an unknown member of Ψ. Evidence E_c is to the effect that the crime stain is of group Γ. This is to say that an unknown member of Ψ is Γ. The probability of this is the probability that a person drawn at random from Ψ is of group Γ which is γ. Thus

$$Pr(E_c | \bar{C}, I) = \gamma.$$

The likelihood ratio V is then

$$V = \frac{Pr(E_c | C, E_s, I)}{Pr(E_c | \bar{C}, I)} = \frac{1}{\gamma}.$$

This value, $1/\gamma$, is the value of the blood grouping evidence when the criminal is a member of Ψ.

6.2.3   Particular population

Suppose now that I does provide information about the criminal, relevant to the blood group frequencies, and the relevant population is now Ψ_0, a subset of Ψ. For example, as mentioned above, I may include an eyewitness description of the criminal as Chinese. Suppose the frequency of blood group Γ amongst Chinese is $100\beta\%$. Then $Pr(E_c | \bar{C}, I) = \beta$ and

$$V = \frac{1}{\beta}.$$

Numerical examples follow in Section 6.2.4.

6.2.4   Examples

Buckleton *et al.* (1987) provide data for racial blood group gene frequencies in New Zealand from which numerical examples applicable to Sections 6.2.2 and

Table 6.1   Gene frequencies for New Zealand in the *ABO* system

Blood group	A	B	O
Relative frequency	0.254	0.063	0.683

Table 6.2   Gene frequencies for Chinese in New Zealand in the *ABO* system

Blood group	A	B	O
Relative frequency	0.168	0.165	0.667

6.2.3 may be derived. Consider the *ABO* blood grouping system and that the source (suspect) and receptor (crime) stain are both of group *B*.

General population

From Buckleton *et al.* (1987), the gene frequencies for New Zealand in this system are given in Table 6.1.

Thus, for a crime for which the relevant population Ψ was the general New Zealand population,

$$V = 1/0.063 = 15.87 \simeq 16.$$

The evidence is 16 times more likely if the suspect was present at the crime scene than if he was not.

Particular population

Background information I has included an eyewitness description of the criminal as Chinese. From Buckleton *et al.* (1987) the gene frequencies for Chinese in the *ABO* system are as given in Table 6.2.

Then for a crime for which the Chinese population, Ψ_0, a subset of Ψ, was the relevant population,

$$V = 1/0.165 = 6.06 \simeq 6.$$

The evidence is six times more likely if the suspect was at the crime scene than if he was not. Thus the value of the evidence has been reduced by a factor of $15.87/6.06 = 2.62$ if there is external evidence that the criminal was Chinese.

Note, however, as discussed in Section 3.1, that in cases regarding transfer to the crime scene by the criminal, evidence regarding blood group frequencies has to relate to the population from which the criminal has come, not that of the suspect (though they may be the same). It is not relevant to an investigation for a suspect to be detained, for it to be noted that he is Chinese and for this to provide the sole ground for the use of blood group frequencies in the Chinese population. In order to use blood group frequencies in the Chinese population I has to contain information, such as eyewitness evidence, about the criminal. However, for evidence regarding transfer from the crime scene to the criminal, the life style of the suspect may be relevant.

Note also, as discussed in Buckleton *et al.* (1987), the blood group frequencies in the general population have been derived from a weighted average of the blood group frequencies in each race or subpopulation which make up the population from which the criminal may be thought to have come. The weights are taken to be the proportion of each race in the general population.

6.3 TWO SAMPLES

6.3.1 Two stains, two offenders

The single sample case described in Section 6.2 can be extended to a case in which two blood stains have been left at the crime scene (Evett, 1987b). A crime has been committed by two men, each of whom left a bloodstain at the crime scene. The stains are grouped; one is found to be of group Γ_1, the other of group Γ_2. Later, as a result of information completely unrelated to the blood evidence, a single suspect is identified. His blood is found to be of group Γ_1. It is assumed there is no evidence in the form of injuries. The scientific evidence is confined solely to the results of the blood grouping. The two hypotheses to be considered are:

- C: the crime stain came from the suspect and one other man;
- $\bar{C}$: the crime stain came from two other men.

The scientific evidence E consists of:

(a) the two crime stains of groups Γ_1 and Γ_2, (E_c),
(b) the suspect's blood is of group Γ_1, (E_s) (without loss of generality since a similar result follows if the suspect is of group Γ_2).

The value, V, of the evidence is given by

$$\frac{Pr(E_c|C, E_s, I)}{Pr(E_c|\bar{C}, I)}. \tag{6.2}$$

Notice that if the blood stain did not come from the suspect his blood group is not relevant. The scientist knows that blood of groups Γ_1 and Γ_2 occur with probabilities γ_1 and γ_2, respectively, in some relevant population.

Assume that *I* contains no information which may restrict the definition of the relevant population to a subset of the general population.

Consider the numerator of (6.2) first. This is the probability that the two stains are of groups Γ_1 and Γ_2, given that

- the suspect is the source of one of the crime stains;
- the suspect is of group Γ_1 and
- all other information, *I*; from the assumption above *I* implies that blood group frequencies in the general population are the relevant ones.

Assume that the two crime stains are independent pieces of evidence in the sense that knowledge of the group of one of the stains does not influence the probability that the other will be of a particular group. Let the two criminals be denoted *A* and *B*.

Also, E_c may be considered in two mutually exclusive partitions.

- E_{c1}: *A* is of group Γ_1, *B* is of group Γ_2,
- E_{c2}: *A* is of group Γ_2, *B* is of group Γ_1.

Partitions E_{c1} and E_{c2} may be further subdivided using the assumption of independence. Thus $E_{c1} = (E_{c11}, E_{c12})$ where

- E_{c11}: *A* is of group Γ_1,
- E_{c12}: *B* is of group Γ_2.

Similarly, $E_{c2} = (E_{c21}, E_{c22})$ where

- E_{c21}: *A* is of group Γ_2,
- E_{c22}: *B* is of group Γ_1.

Thus, since E_{c1} and E_{c2} are mutually exclusive:

$$Pr(E_c|C, E_s, I) = Pr(E_{c1}, E_{c2}|C, E_s, I)$$
$$= Pr(E_{c1}|C, E_s, I) + Pr(E_{c2}|C, E_s, I),$$

(from (1.2), the second law of probability for mutually exclusive events). However, only one of these two probabilities is non-zero. If the suspect is *A* then the latter probability is zero; if the suspect is *B* then the former probability is zero. Assume, again without loss of generality, that the suspect is *A*. Then

$$Pr(E_c|C, E_s, I) = Pr(E_{c1}|C, E_s, I)$$
$$= Pr(E_{c11}|C, E_s, I) \times Pr(E_{c12}|C, E_s, I),$$

by independence (1.6).

Now, $Pr(E_{c11}|C, E_s, I) = 1$ since, if the suspect was the source of one of the crime stains and his blood group is Γ_1 then it is certain that one of the crime stains is Γ_1.

Also, $Pr(E_{c12}|C, E_s, I) = \gamma_2$ since the second blood stain was left by the other criminal. At present, in the absence of information from I, B is considered as a member of the relevant population. The probability is thus the relative frequency of group Γ_2 in the relevant population and this is γ_2. The numerator then takes the value γ_2.

Now consider the denominator of (6.2):

$$Pr(E_c|\bar{C}, I) = Pr(E_{c1}|\bar{C}, I) + Pr(E_{c2}|\bar{C}, I)$$
$$= Pr(E_{c11}|\bar{C}, I)Pr(E_{c12}|\bar{C}, I) + Pr(E_{c21}|\bar{C}, I)Pr(E_{c22}|\bar{C}, I),$$
$$= \gamma_1\gamma_2 + \gamma_2\gamma_1$$
$$= 2\gamma_1\gamma_2.$$

Thus,
$$V = \gamma_2/(2\gamma_1\gamma_2) = 1/(2\gamma_1). \tag{6.3}$$

This result should be compared with the result for the single sample case, $V = 1/\gamma$. The likelihood ratio in the two-sample case is one half of what it is in the corresponding single sample case. This is intuitively reasonable. If there are two criminals and one suspect, one would not expect the evidence of a matching blood stain to be as valuable as in the case in which there is one criminal and one suspect.

6.3.2   DNA profiling

The evidence from a DNA profile may be interpreted simplistically as discrete data by the use of relative frequencies, determined from an appropriate histogram, rather than probability density estimates, in the assessment of the evidential value of the band positions. Other approaches based on continuous data are described in Chapter 8. An application of this interpretation to DNA profiles from single locus probes from more than one person is described by Evett *et al.* (1991).

A woman has claimed to have been raped by two men. A suspect A has been arrested. A sample from a vaginal swab provides, using a single locus probe, evidence E of four bands, in positions w, x, y and z. The suspect's sample provides bands in positions x and y, matching two of the bands in the crime sample. Neither the exact criteria for determining a match, nor the method for determining a band frequency are at issue here. Both of these issues will be discussed later in Chapter 8. Assume the relative frequencies of the band positions of w, x, y and z in the general population are p_w, p_x, p_y and p_z, respectively. It is assumed that none of the bands in the crime sample can be attributed to the victim.

The two hypotheses to be considered in the evaluation of the likelihood ratio are:

- C: the crime sample came from the suspect A and one other man;
- $\bar{C}$: the crime sample came from two other men.

If C were true, the suspect has contributed bands x and y with probability 1. The probability that the other man would contribute bands w and z is $2p_w p_z$. Thus

$$Pr(E|C) = 2p_w p_z.$$

If $\bar{C}$ were true there are six $\left\{ \binom{4}{2} \right\}$ ways in which the two men could contribute bands to make up the observed crime profile. Each way has probability $4p_w p_x p_y p_z$. Thus,

$$Pr(E|\bar{C}) = 24 p_w p_x p_y p_z.$$

The value of the evidence is thus

$$V = \frac{Pr(E|C)}{Pr(E|\bar{C})} = \frac{2p_w p_z}{24 p_w p_x p_y p_z} = \frac{1}{12 p_x p_y}.$$

If there had been only one assailant and only bands in positions x and y present then this likelihood ratio would have the value $1/(2p_x p_y)$. The existence of the other man has reduced the value of the evidence by a factor of 6.

Other examples are given by Evett *et al.* (1991). Suppose another suspect, B, provides a sample which gives a profile consisting of bands in positions w and z. If the two hyotheses are now:

- C: the crime sample came from suspects A and B,
- $\bar{C}$: the crime sample came from two unknown men,

then

$$V = \frac{1}{24 p_w p_x p_y p_z},$$

since the numerator has the value 1.

Further examples include the situation in which the single locus probe analysis reveals only three bands, in positions w, x and y, not four. The two hypotheses considered are:

- C: the crime sample came from A and one other man,
- $\bar{C}$: the crime sample came from two other men.

Assume the band in position w is a single band and the conservative estimate $2p_w$

(see Section 8.2.2) is used for its frequency. Then

$$V = \frac{(1 + p_x + p_y)}{4p_x p_y (3 + p_w + p_y + p_z)}.$$

If the estimate p_w^2 is used for its frequency, then

$$V = \frac{(p_w + 2p_x + 2p_y)}{12 p_x p_y (p_w + p_x + p_y)}.$$

6.4   MANY SAMPLES

Consider a crime in which n blood stains are left at the crime scene. A single suspect is identified whose blood type matches that of one of the blood stains at the crime scene. Assume throughout that I contains no relevant information. These specific examples, while hypothetical, illustrate points which require consideration in the evaluation of evidence.

6.4.1   Many different blood types

Assume the crime sample consists of n blood stains, all of different blood groups, one from each of n offenders. The two hypotheses to be considered are:

- C: the crime sample came from the suspect and $(n-1)$ other men;
- $\bar{C}$: the crime sample came from n unknown men.

The scientific evidence E consists of:

- E_c: the crime stains are of groups $\Gamma_1, \Gamma_2, \ldots, \Gamma_n$,
- E_s: the suspect's blood group is of group Γ_1 (without loss of generality).

Consider the numerator $Pr(E_c | C, E_s, I)$. The suspect's blood group matches that of the stain of group Γ_1. There are $(n-1)$ other criminals who can be allocated to the $(n-1)$ other stains in $(n-1)!$ ways. Thus:

$$Pr(E_c | C, E_s, I) = (n-1)! \prod_{i=2}^{n} \gamma_i = (n-1)!(\gamma_2, \ldots, \gamma_n).$$

Now, consider the denominator. There are $n!$ ways in which the n criminals, of whom the suspect is not one, can be allocated to the n stains. Thus:

$$Pr(E_c | \bar{C}, I) = ! \prod_{i=1}^{n} \gamma_2 = n!(\gamma_1, \ldots, \gamma_n).$$

Hence

$$V = \frac{(n-1)!\prod_{i=2}^{n}\gamma_i}{n!\prod_{i=1}^{n}\gamma_i} = \frac{1}{n\gamma_1}. \tag{6.4}$$

6.4.2 General cases

n stains, k groups, k offenders

Suppose now that there are k different blood groups and that these correspond to k different people among the n stains $(k < n)$ forming the crime sample and that the suspect has one of these blood groups. The two hypotheses to be considered are:

- C: the crime sample came from the suspect and $(k-1)$ other men;
- $\bar{C}$: the crime sample came from k unknown men.

The scientific evidence consists of:

- E_c: the crime stains are of groups $\Gamma_1, \ldots, \Gamma_k$ and there are $s_1, \ldots, s_k (\sum_{i=1}^{k} s_i = n)$ of each,
- E_s: the suspect's blood is of group Γ_1 (without loss of generality).

The probabilities given below are in the form of the multinomial distribution, a generalisation of the binomial distribution of Section 3.3.3.

Consider the numerator $Pr(E_c|C, E_s, I)$. The suspect's blood group matches that of the stains of group Γ_1. There are $(n - s_1)$ other blood stains which can be allocated in $(n - s_1)!/(s_2! \cdots s_k!)$ ways to give

$$Pr(E_c|C, E_s, I) = \frac{(n-s_1)!}{s_2! \cdots s_k!} \gamma_2^{s_2} \cdots \gamma_k^{s_k}.$$

Now, consider the denominator. There are $n!/(s_1!s_2!\cdots s_k!)$ ways in which the n stains, none of which is associated with the suspect, can be allocated to the blood groups. Thus

$$Pr(E_c|\bar{C}, I) = \frac{n!}{s_1!\cdots s_k!} \gamma_1^{s_1} \cdots \gamma_k^{s_k}.$$

Hence

$$V = \frac{(n-s_1)!s_1!}{n!\gamma_1^{s_1}} = \frac{1}{\binom{n}{s_1}\gamma_1^{s_1}},$$

where $\binom{n}{s_1}$ is the binomial coefficient (3.2). Notice that V is independent of k, the number of criminals and that if $s_1 = 1$ the result reduces to that of (6.4).

n stains, k groups, m offenders

A similar result may be obtained in the following situation. There are n bloodstains with k different groups with s_i in the i-th group ($\sum_{i=1}^{k} s_i = n$). There are m offenders with m_i in each blood group ($\sum_{i=1}^{k} m_i = m$) such that $s_{ij}(j = 1, \ldots, m_i)$ denotes the number of stains belonging to the j-th offender in the i-th group. The denominator equals

$$\frac{n!}{s_{11}! \cdots s_{km_k}!} \gamma_1^{s_1} \cdots \gamma_k^{s_k}.$$

The numerator equals

$$\frac{(n - s_{11})!}{s_{12}! \cdots s_{km_k}!} \gamma_1^{s_1 - s_{11}} \gamma_2^{s_k} \cdots \gamma_k^{s_k}.$$

when it is assumed, without loss of generality, that the first set of stains in the first group came from the suspect. Then

$$V = \frac{(n - s_{11})! \; s_{11}!}{n!} \times \frac{1}{\gamma_1^{s_{11}}} = \frac{1}{(s_{11}^n) \gamma_1^{s_{11}}}.$$

6.5 RELEVANCE OF EVIDENCE AND RELEVANT MATERIAL

6.5.1 Introduction

An extension of the results of the previous section to deal with two further issues is considered by Evett (1993a). The first issue concerns material that may not be *relevant* (Stoney, 1991 and 1994). Crime material which came from the offender is said to be relevant in that it is relevant to the consideration of suspects as possible offenders; see Section 5.3.1 for a brief discussion of this idea. Relevant material as discussed here should be distinguished from relevant populations as defined in Section 5.4. The second issue concerns the recognition that if the material is not relevant to the case then it may have arrived at the scene from the suspect for innocent reasons. In this section reference to I has, in general, been omitted for clarity. A crime has been committed by k offenders. A single blood stain is found at the crime scene in a position where it may have been left by one of the offenders. A suspect is found and he gives a blood sample. The suspect's sample and the crime stain are of the same blood group Γ with frequency γ amongst the relevant population from which the criminals have come. As before, consider two hypotheses

- C: the suspect is one of the k offenders,
- $\bar{C}$: the suspect is not one of the k offenders.

Notice the difference between these hypotheses and those of Section 6.4. There, the hypotheses referred to the suspect being, or not being, one of the donors of the blood stains found at the crime scene. Now, the hypotheses are stronger, namely that the suspect is, or is not, one of the offenders. The value V of the evidence is

$$V = \frac{Pr(E_c|C, E_s)}{Pr(E_c|\bar{C})} \tag{6.5}$$

where E_c is the blood group Γ of the crime stain and E_s is the blood group Γ of the suspect.

6.5.2   Association hypotheses

The introduction of hypotheses known as *association hypotheses* has been suggested by Buckleton (personal communication, cited by Evett, 1993a). First, consider the following:

- B: the crime stain came from one of the k offenders,
- $\bar{B}$: the crime stain did not come from any of the k offenders.

The value, V, of the evidence may now be written using the law of total probability (Section 1.6.6) as

$$V = \frac{Pr(E_c|C, B, E_s)Pr(B|C, E_s) + Pr(E_c|C, \bar{B}, E_s)Pr(\bar{B}|C, E_s)}{Pr(E_c|\bar{C}, B)Pr(B|\bar{C}) + Pr(E_c|\bar{C}, \bar{B})Pr(\bar{B}|\bar{C})}.$$

In the absence of E_c, the evidence of the blood group of the crime stain, knowledge of C and of E_s does not effect our belief in the truth or otherwise of B. This is what is meant by *relevance* in this context. Thus

$$Pr(B|C, E_s) = Pr(B|C) = Pr(B)$$

and

$$Pr(\bar{B}|\bar{C}, E_s) = Pr(\bar{B}|\bar{C}) = Pr(\bar{B}).$$

Let $Pr(B) = r$, $Pr(\bar{B}) = (1 - r)$ and call r the *relevance term*, i.e., relevance is equated to the probability the stain had been left by one of the offenders. The higher the value of r, the more relevant the stain becomes. Thus

$$V = \frac{Pr(E_c|C, B, E_s)r + Pr(E_c|C, \bar{B}, E_s)(1 - r)}{Pr(E_c|\bar{C}, B, E_s)r + Pr(E_c|\bar{C}\bar{B}, E_s)(1 - r)}.$$

6.5.3   Intermediate association hypotheses

In order to determine the component probabilities of this expression two further hypotheses known as *intermediate association hypotheses* are introduced:

- A: the crime stain came from the suspect,
- $\bar{A}$: the crime stain did not come from the suspect.

Now consider the four conditional probabilities above.

$Pr(E_c|C, B, E_s)$

This is the probability that the crime stain would be of group Γ if it had been left by one of the offenders (B), the suspect had committed the crime (C) and the suspect is of blood group Γ:

$$Pr(E_c|C, B, E_s) = Pr(E_c|C, B, A, E_s)Pr(A|C, B, E_s) + Pr(E_c|C, B, \bar{A}, E_s)Pr(\bar{A}|C, B,E_s).$$

Here $E_c = E_s = \Gamma$ and $Pr(E_c|C, B, A, E_s) = 1$. In the absence of E_c, A is independent of E_s and so

$$Pr(A|C, B, E_s) = Pr(A|C, B) = 1/k,$$

where it has been assumed that there is nothing in the background information I to distinguish the suspect, given C, from the other offenders as far as blood shedding is considered.

In a similar manner, $Pr(\bar{A}|C, B, E_s) = (k - 1)/k$. Also,

$$Pr(E_c|E_s, C, B, \bar{A}) = Pr(E_c|C, B, \bar{A}) = \gamma,$$

since if $\bar{A}$ is the case E_c and E_s are independent and one of the other offenders left the stain (since B holds). Thus

$$Pr(E_c|C, B, E_s) = \{1 + (k - 1)\gamma\}/k.$$

$Pr(E_c|C, \bar{B}, E_s)$

This is the probability that the crime stain would be of group Γ if it had been left by an unknown person who was unconnected with the crime. (This is the implication of assuming $\bar{B}$ to be true.) The population of people who may have left the stain is not necessarily the same as the population from which the criminals are assumed to have come. Thus, let

$$Pr(E_c|C, \bar{B}, E_s) = \gamma'.$$

where γ' is the probability of blood group Γ amongst the population of people who may have left the stain (the prime ' indicating it may not be the same value as γ which relates to the population from which the criminals have come).

Innocent suspect, $\bar{C}$ is true

First, note that

$$Pr(E_c|\bar{C}, B, E_s) = Pr(E_c|\bar{C}, B) = \gamma,$$

the frequency of Γ amongst the population from which the criminals have come. Also,

$$Pr(E_c|\bar{C}, \bar{B}, E_s) = Pr(E_c|\bar{C}, \bar{B}, A, E_s)Pr(A|\bar{C}, \bar{B}, E_s)$$
$$+ Pr(E_c|\bar{C}, \bar{B}, \bar{A}, E_s)Pr(\bar{A}|\bar{C}, \bar{B}, E_s).$$

If A is the case, $Pr(E_c|\bar{C}, \bar{B}, A, E_s) = 1$. Also $Pr(A|\bar{C}, \bar{B}, E_s) = Pr(A|\bar{C}, \bar{B})$.This is the probability that the stain would have been left by the suspect even though he was innocent of the offence. Denote this probability by p. Here it is assumed that the propensity to leave a stain is independent of the blood group of the person leaving the stain. Hence $Pr(A|\bar{C}, \bar{B}) = p$ and $Pr(\bar{A}|\bar{C}, \bar{B}, E_s) = Pr(\bar{A}|\bar{C}, \bar{B}) = 1 - p$. Also $Pr(E_c|\bar{C}, \bar{B}, \bar{A}) = \gamma'$. Thus

$$Pr(E_c|\bar{C}, \bar{B}, E_s) = p + (1 - p)\gamma'.$$

Substitution of the above expressions into (6.5) gives

$$V = \frac{[r\{1 + (k-1)\gamma\}/k] + \{\gamma'(1-r)\}}{\gamma r + \{p + (1-p)\gamma'\}(1-r)}$$
$$= \frac{r\{1 + (k-1)\gamma\} + k\gamma'(1-r)}{k[\gamma r + \{p + (1-p)\gamma'\}(1-r)]}.$$

6.5.4   Examples

Example 6.1   Consider the case where it may be assumed that γ and γ' are approximately equal and that $p = 0$. The latter assumption holds if there is no possibility that the suspect may have left the stain for innocent reasons. Then

$$V = \frac{r\{1 + (k-1)\gamma\} + k\gamma(1-r)}{k\{\gamma r + \gamma(1-r)\}}$$
$$= \frac{r + (k-r)\gamma}{k\gamma}. \tag{6.6}$$

If γ is so small that $r/k\gamma \gg 1$ then $V \simeq r/k\gamma$. If $r = 1$, $V \simeq 1/k\gamma$, see (6.4). This is a reasonable result, the value of the evidence has been reduced by a factor corresponding to the number of offenders.

Example 6.2   Assume $p \neq 0$ but that γ and γ' are approximately equal. Then

$$V = \frac{r + (k - r)\gamma}{k[p(1 - r) + \gamma\{r + (1 - p)(1 - r)\}]}$$

$$= \frac{r + (k - r)\gamma}{k[p(1 - r) + \gamma\{1 - p + pr\}]}.$$

Subjective probabilities

At this stage it is necessary to think more about r and p. The first, r, is the probability that the crime stain came from one of the offenders, and this probability has been defined as the *relevance* of the crime stain. The second, p, is the probability the crime stain came from the suspect, given the suspect did not commit the crime and that the crime stain did not come from any of the offenders; i.e., it is the probability the crime stain was left there innocently by someone who is now a suspect. The validity of combining probabilities, thought of as measures of belief, and probabilities as relative frequencies has been questioned (Freeling and Sahlin, 1983; Stoney, 1994). However, the discussion in Section 1.6.1 and comments on *duality* by Hacking (1975) explain why such combinations can be valid.

Evett (1993a) suggests that determination of the probabilities such as those above may be the province of the court and that it is necessary to establish the conditions under which the scientific evidence can be of any guidance to the court. Evett suggests an examination of the sensitivity of V to values of p and of r. As an illustration, he takes the number of offenders k to be 4 and the frequencies γ and γ' to be 0.001. Then

$$V = \frac{r + 0.004}{4[p(1 - r) + 0.001(1 - p + pr)]}$$

where $r + (k - r)\gamma$ has been approximated by $r + k\gamma$.

The variation of V with r and p is shown in Figure 6.1. The graph has been drawn with a logarithmic scale for V. This is plotted against p for $r = 1$, 0.75, 0.50, 0.25. It is useful to consider individually the terms within the expression for V for the case in which there is one blood stain of group Γ and frequency γ (which is assumed to be well known):

- The number of offenders, k: this is assumed to be well known.
- Relevance, r: the (subjective) probability that the crime stain came from one of the offenders; factors to be considered in its estimation include location, abundance and apparent freshness of the blood.

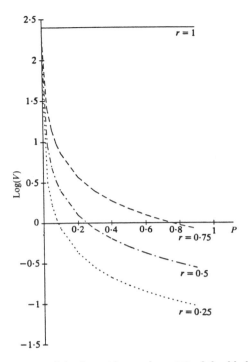

Figure 6.1 The variation of the logarithm to base 10 of the likelihood ratio V of the evidence with p, the probability that the stain would have been left by the suspect even though he was innocent of the offence, for various values of r, the probability that the stain would have been left by one of the offenders. The number of offenders, k, equals 4 and the relative frequency γ of the blood group of the crime stain is 0.001 (Evett, 1993a). (Reproduced with permission of The Forensic Science Society.)

- Innocence, p: the (subjective) probability that the crime stain came from the suspect, given the suspect did not commit the crime and the crime stain did not come from any of the offenders.

Appropriate values for the probabilities of relevance and innocence are matters for the courts to decide.

In general, V decreases as r decreases or as p increases.

For $r = 1$, it is certain that the the crime stain came from one of the offenders and

$$V = \frac{1.004}{0.004},$$

$$= 251,$$

$$\log_{10}(V) = 2.40.$$

For $r \neq 1$, V is very sensitive to p. If there is a non-zero probability that the crime stain did not come from one of the offenders then the probability of innocence has a considerable influence on V. For example, if $r = 0.25$, so that there is a small probability the crime stain came from one of the offenders, V becomes less than 1 for $p > 0.083 \simeq 1/12$. Thus, if $p > 1/12$ (and there is a small probability that the crime stain came from the suspect, conditional on everything else) then the evidence supports the hypothesis that some other person is the offender rather than the hypothesis that the suspect is the offender.

6.5.5 Two stains, one offender

The two-trace bloodstain problem of Section 6.3.1 has been modified by Stoney (1994) to the case where there are two bloodstains, of group Γ_1 and Γ_2, respectively, as before but only one offender (rather than two as in Section 6.3.1) who left one of the bloodstains but it is not known which one. A suspect is found who is of group Γ_1. Relevance is applicable in this context since it provides a measure of belief (probability) that the stain at the crime scene which comes from the offender is the one which is of the same group as the suspect. The two competing hypotheses to be considered are:

- C: the suspect is the offender;
- $\bar{C}$: the suspect is not the offender.

Let r be the probability that the matching stain (Γ_1) was from the offender. As before this is a subjective probability. It has been assumed that one of the stains is from the offender so there is a probability $(1 - r)$ that it is the other stain (Γ_2) that is from the offender.

Suppose C is true. Then

$$Pr(\text{suspect's group and crime group correspond} \mid$$
$$\text{stain } \Gamma_1 \text{ was from the offender}) = 1$$
$$Pr(\text{stain } \Gamma_1 \text{ was from the offender}) = r.$$
$$Pr(\text{suspect's group and crime group correspond} \mid$$
$$\text{stain } \Gamma_2 \text{ was from the offender}) = 0.$$
$$Pr(\text{stain } \Gamma_2 \text{ was from the offender}) = (1 - r).$$

Thus $Pr(\text{match in } \Gamma_1 \mid C) = r$.

Suppose $\bar{C}$ is true. The probability of a correspondence in Γ_1 is the probability that a randomly selected person is of type Γ_1. This is the phenotype frequency γ_1. The likelihood ratio is then

$$V = r/\gamma_1.$$

This is a particular case of (6.6) with small γ_1 and $k = 1$.

If the two stains have equal probabilities of being left by the offender, $r = 1/2$ and the likelihood ratio equals $1/2\gamma_1$. This is numerically equivalent to the figure $1/2\gamma_1$ derived by Evett (1987b) and quoted above in (6.3) for a problem with two stains, one left by each of the two offenders, and a single suspect whose blood group matches that of one of the stains and whose frequency is γ_1. The derivation, however, is very different.

Stoney (1994) continued the development to the case where neither stain may be relevant but there is still a single offender. The suspect has blood group Γ_1. Let there be the following probabilities:

$$Pr(\text{The stain of group } \Gamma_1 \text{ is from the offender}) = r_1,$$
$$Pr(\text{The stain of group } \Gamma_2 \text{ is from the offender}) = r_2,$$
$$Pr(\text{Neither stain is from the offender}) = 1 - r_1 - r_2.$$

If C is true, there are three components to the probability:

- Stain of type Γ_1 is from the offender. There is a match with probability r_1.
- Stain of type Γ_2 is from the offender. There is no match. This event has probability zero since the suspect is assumed to be the offender and only one offender is assumed.
- Neither stain is from the offender. This event has probability $(1 - r_1 - r_2)$ and if it is true there is a probability γ_1 of a match between the suspect's blood group (Γ_1) and the crime stain of the same group. The probability of the combination of these events is $(1 - r_1 - r_2)\gamma_1$.

These three components are mutually exclusive and the probability in the numerator of the likelihood ratio is the sum of these three probabilities, namely $r_1 + (1 - r_1 - r_2)\gamma_1$. If $\bar{C}$ is true, the probability of a correspondence is as before, namely γ_1.

The likelihood ratio is then

$$V = \frac{r_1 + (1 - r_1 - r_2)\gamma_1}{\gamma_1}.$$

Certain special cases can be distinguished. As r_1 and r_2 tend to zero, which implies that neither stain is relevant, then the likelihood ratio tends to 1. A likelihood ratio of 1 provides no support for either hypothesis, a result in this case which is entirely consistent with the information that neither stain is relevant. For $r_1 = r_2 = 1/2$, $V = 1/2\gamma_1$. For $r_1 = 1$, $V = 1/\gamma_1$. As $r_2 \to 1$, then $r_1 \to 0$ and $V \to 0$. All of these are perfectly reasonable results.

6.6 SUMMARY

The results of the previous sections relating to blood stains may usefully be summarised.

6.6.1   Stain known to have been left by offenders

One stain known to have come from one offender

The blood group of the crime stain and of the suspect is Γ with frequency γ. The hypotheses to be compared are:

- C: there was contact between the suspect and the crime scene;
- $\bar{C}$: there was not contact between the suspect and the crime scene.

Then

$$V = \frac{1}{\gamma}.$$

Two stains, one from each of two offenders

There are two crime stains, of groups Γ_1 and Γ_2 with frequencies γ_1 and γ_2. There is one suspect with blood group Γ_1 with frequency γ_1. The hypotheses to be compared are:

- C: the crime stain came from the suspect and one other man;
- $\bar{C}$: the crime stains came from two unknown men.

Then

$$V = \frac{1}{2\gamma_1}.$$

n stains, one from each of n offenders

There is one suspect with blood group Γ_1 with frequency γ_1. The hypotheses to be compared are

- C: the crime sample came from the suspect and $(n-1)$ other men;
- $\bar{C}$: the crime sample came from n unknown men.

Then

$$V = \frac{1}{n\gamma_1}.$$

n stains, k groups, k offenders

There are s_i stains of type $i (i = 1, \ldots, k; \sum_{i=1}^{k} s_i = n)$. There is one suspect with blood group Γ_1 with frequency γ_1. The hypotheses to be compared are

- C: the crime sample came from the suspect and $(k-1)$ other men;
- $\bar{C}$: the crime sample came from k unknown men.

Then

$$V = \frac{1}{\binom{n}{s_1} \gamma_1^{s_1}}.$$

n stains, k groups, m offenders

There are m_i offenders in group i ($i = 1, \ldots, k$; $\sum_{i=1}^{k} m_i = m$). There are s_{ij} stains belonging to the j-th offender in the i-th group. There is one suspect with blood group Γ_1 with frequency γ_1. Assume this could be the first offender in the first group. The hypotheses to be compared are:

- C: the crime sample came from the suspect and ($m - 1$) other men;
- $\bar{C}$: the crime sample came from m other men.

Then

$$V = \frac{1}{\binom{n}{s_{11}} \gamma_1^{s_{11}}}.$$

6.6.2 Relevance—stain may not have been left by offenders

One stain, k offenders

Relevance: the probability that a crime stain came from one of the k offenders is r. The hypotheses to be compared are

- C: the suspect is one of the k offenders;
- $\bar{C}$: the suspect is not one of the k offenders.

The stain is of group Γ. It may have been left by an offender. There are k offenders. The suspect is of group Γ. This group has frequency γ in the population from which the criminals may be thought to come. It has frequency γ' amongst the population of people who may have left the stain which may not be the same population as that from which the criminals may be thought to come. For example, there may be eyewitness evidence that the criminals are from one ethnic group whereas the people normally associated with the crime scene may be from another. The probability that the stain would have been left by the suspect even though he was innocent of the offence is p:

$$V = \frac{[r\{1 + (k-1)\gamma\}/k] + \{\gamma'(1-r)\}}{\gamma r + \{p + (1-p)\gamma'\}(1-r)}$$

$$= \frac{r\{1 + (k-1)\gamma\} + k\gamma'(1-r)}{k[\gamma r + \{p + (1-p)\gamma'\}(1-r)]}.$$

There are several simplifications.

If $\gamma = \gamma'$ and $p = 0$ then

$$V = \frac{r + (k - r)\gamma}{k\gamma}.$$

If, also, $r = 1$,

$$V = \frac{1 + (k - 1)\gamma}{k\gamma}.$$

Compare this with the case in which there are n stains (rather than 1) and the number (k) of offenders equals the number of stains (n). (There is one stain from each of n offenders.) Then $V = 1/n\gamma = 1/k\gamma$.

If there is one stain and k offenders

$$V = \frac{1 + (k - 1)\gamma}{k\gamma}$$

$$= 1 + \frac{1}{k\gamma} - \frac{1}{k},$$

an increase from $1/k\gamma$. However, if γ is small, V is approximately equal to $1/k\gamma$. The value of the evidence is the same when there are k offenders and one suspect whether there is one stain, matching the blood group of the suspect, or many stains of different blood groups.

If $\gamma = \gamma'$ and $p \neq 0$ then

$$V = \frac{[r\{1 + (k - 1)\gamma\}/k] + \{\gamma(1 - r)\}}{k[\gamma r + \{p + (1 - p)\gamma\}(1 - r)]}$$

$$= \frac{r + (k - r)\gamma}{k[p(1 - r) + \gamma\{r + (1 - p)(1 - r)\}]}$$

$$= \frac{r + (k - r)\gamma}{k[p(1 - r) + \gamma(1 - p + pr)]}.$$

Two stains, one of which is relevant, one offender

The offender left one of the blood stains but it is not known which one. The hypotheses to be compared are:

- ˜C: the suspect is the offender,
- $\bar{C}$: the suspect is not the offender.

A suspect is of group Γ_1, with frequency γ_1. Let r be the probability that the crime stain which matches the group of the suspect is from the suspect. Then $(1 - r)$ is the probability that the crime stain which does not match the blood group of the suspect is from the offender. Then

$$V = \frac{r}{\gamma_1}.$$

Two stains, neither of which may be relevant, one offender

The hypotheses to be compared are:

- C: the suspect is the offender,
- $\bar{C}$: the suspect is not the offender.

A suspect is of group Γ_1, with frequency γ_1. Let r_1 and r_2 be the probabilities that the stain of group Γ_1 and the stain of group Γ_2 is from the offender, respectively. Then $(1 - r_1 - r_2)$ is the probability that neither stain is from the offender and

$$V = \frac{r_1 + (1 - r_1 - r_2)\gamma_1}{\gamma_1}.$$

6.7 MISSING PERSONS

Consider a case of a missing person in which there is evidence of foul play and a suspect has been identified. Evidence of the missing person's blood group and other phenotypic values is not available. Instead, values are available from the parents of the missing person (case couple) from which inferences about the missing person may be made. This is parentage testing; see Section 5.1.3. Blood stains found on the suspect's property could have come from one of the parent's offspring. The hypotheses to be compared are:

- C: the blood stains came from a child of the case couple;
- $\bar{C}$: the blood stains did not come from a child of the case couple.

The value of the evidence is

$$V = \frac{\text{Probability parents would pass the stain phenotype}}{\text{Probability a random couple would pass the stain phenotype}}.$$

This may be expressed verbally as 'the parents of the missing person are V times more likely than a randomly selected couple to pass this set of genes'.

Consider the following two cases of missing persons with phenotypes given in Table 6.3, the first (Case 1) described by Kuo (1982), the second (Case 2) by Ogino and Gregonis (1981), both reviewed by Stoney(1984).

Table 6.3 Phenotypes for two cases of missing persons

Marker System	Case 1			Case 2		
	Father	Mother	Stain	Father	Mother	Stain
ABO	B	O	B	O	O	O
EAP	BA	BA	A	B	BA	B
AK	1	2–1	2–1	1	1	1
ADA	1	1	1	1	1	1
PGM	2–1	1	2–1	1	1	1
Hp	2–1	2	2–1	1	1	1
EsD				1	1	1

Table 6.4 Gene frequencies and phenotypic incidences for Case 1 (Kuo, 1982)

Marker system	Stain phenotype	Gene frequencies	Phenotypic incidences
ABO	B	B: 0.074	$0.074^2 +$
		O: 0.664	$2 \times 0.074 \times 0.664 = 0.1040$
EAP	A	A: 0.327	$0.327^2 \qquad = 0.1069$
AK	2–1	2: 0.038	
		1: 0.962	$2 \times 0.038 \times 0.962 = 0.0731$
ADA	1	1: 0.952	$0.952^2 \qquad = 0.9063$
PGM	2–1	2: 0.229	
		1: 0.771	$2 \times 0.229 \times 0.771 = 0.3531$
Hp	2–1	2: 0.578	
		1: 0.422	$2 \times 0.578 \times 0.422 = 0.4878$

Table 6.5 Gene frequencies and phenotypic incidences for Case 2 (Ogino and Gregonis, 1981)

Marker system	Stain phenotype	Gene frequencies	Phenotypic incidences
ABO	O	O: 0.664	$0.664^2 = 0.4409$
EAP	B	B: 0.612	$0.612^2 = 0.3745$
AK	1	1: 0.962	$0.962^2 = 0.9254$
ADA	1	1: 0.952	$0.952^2 = 0.9063$
PGM	1	1: 0.771	$0.771^2 = 0.5944$
Hp	1	1: 0.422	$0.422^2 = 0.1781$
EsD	1	1: 0.884	$0.884^2 = 0.7815$

The stain phenotypic frequencies may be calculated for each case. These are the probabilities that a random couple would pass the corresponding phenotype and are given in Tables 6.4 and 6.5.

6.7.1 Case 1: Kuo (1982)

A young woman was missing after going on a boat trip with her boyfriend. He became the main suspect. Bloodstains were found on his boat but there was no known blood of the missing woman to group for comparison with the bloodstains. The blood of the parents of the woman was grouped to try and link the bloodstain on the boat to the missing woman.

The combined phenotypic incidence $= 0.1040 \times 0.1069 \times 0.0731 \times 0.9063 \times 0.3531 \times 0.4878 = 1.2656 \times 10^{-4}$, the multiplication being justified by the independence of the marker systems. Notice the implicit extension of the approach based on the likelihood ratio to include evidence from more than one marker; see Section 5.1.3.

6.7.2 Case 2: Ogino and Gregonis (1981)

A man was reported missing by his family. A suspect, driving the victim's vehicle, was arrested for suspicion of murder. Bloodstains were found in various parts of the vehicle. Blood samples were obtained from the victim's mother and father, as well as other relatives. The combined phenotypic incidence based on those of the parents (Table 6.3), from Table 6.5 is $0.4409 \times 0.3745 \times 0.9254 \times 0.9063 \times 0.5944 \times 0.1781 \times 0.7815 = = 1.146 \times 10^{-2}$.

6.7.3 Calculation of the likelihood ratio

The likelihood ratio V compares two probabilities. The probability parents would pass on a stain phenotype is compared with the probability that a random couple would pass a stain phenotype. The latter is simply the product of the gene frequencies and these values have already been calculated for each of the two cases.

The probability that the parents would pass the stain phenotype is calculated for each case as follows.

The possible couple/stain combinations for up to three codominant alleles are given in Table 6.6.

The *ABO* system requires a special treatment because of dominance. The probabilities of passage depend on the frequencies with which homozygous and heterozygous individuals occur; see Table 6.7. Type *A* and type *B* parents may be either homozygous with the dominant allele or heterozygous with the recessive *O* allele. Further details are contained in Stoney (1984). These frequencies may be calculated directly from the gene frequencies.

Table 6.6   Frequencies for up to three codominant alleles

Couple	Possible stains and frequencies		
PP, PP	$PP = 1.00$		
PP, PQ	$PP = 0.50$	$PQ = 0.50$	
PP, QQ	$PQ = 1.00$		
PQ, PQ	$PP = 0.25$	$PQ = 0.50$	$QQ = 0.25$
PQ, PR	$PP = 0.25$	$PR = 0.25$	$PQ = 0.25$
	$QR = 0.25$		
PQ, RR	$PR = 0.50$	$QR = 0.50$	

Table 6.7   Relative frequencies for *ABO* system

Type *P*	Homozygous *PP*	Heterozygous *PO*
A	$(a)/(a + 2o) = 0.262/1.590 = 0.1648$	$(2o)/(a + 2o) = 1.328/1.590 = 0.8352$
B	$(b)/(b + 2o) = 0.074/1.402 = 0.0528$	$(2o)/(b + 2o) = 1.328/1.402 = 0.9472$

The probability that a particular gene will be passed is then determined by combining the likelihoods of heterozygosity and homozygosity with the likelihood of passing the gene in each instance.

Case 1

The probability that the couple would pass the stain phenotype equals $0.5264 \times 0.25 \times 0.50 \times 1.00 \times 0.50 \times 0.50 = 0.0164$; see Table 6.8. The probability that the stain phenotype occurs by chance equals 1.2656×10^{-4}. Thus

$$V = \frac{0.0164}{1.2656 \times 10^{-4}} = 130.0.$$

The parents of the missing person in Case 1 may be said to be 130 times more likely to pass the stain phenotype than would be a randomly selected couple. From the qualitative scale of Table 2.9 the evidence provides a great increase in support for the hypothesis that the blood stains came from a child of the case couple.

Case 2

The probability that the couple would pass the stain phenotype is 0.50; see Table 6.9. The probability that the stain phenotype occurs by chance $= 1.146 \times 10^{-2}$ and

$$V = \frac{0.50}{1.146 \times 10^{-2}} = 43.6.$$

Table 6.8 The probabilities that the parents will pass the specified stain phenotype; Case 1

Marker system	Couple phenotypes F	M	Stain phenotype	Probability that couple would pass stain phenotype
ABO	B	O	B	0.5264[a]
EAP	BA	BA	A	0.25
AK	1	2–1	2–1	0.50
ADA	1	1	1	1.00
PGM	2–1	1	2–1	0.50
Hp	2–1	2	2–1	0.50

[a] $(1 \times 0.528) + (0.5 \times 0.9472)$

Table 6.9 Probabilities that the parents will pass the specified phenotypes; Case 2

Marker system	Couple phenotypes F	M	Stain phenotype	Probability that couple would pass stain phenotype
ABO	O	O	O	1.00
EAP	B	BA	B	0.50
AK	1	1	1	1.00
ADA	1	1	1	1.00
PGM	1	1	1	1.00
Hp	1	1	1	1.00
EsD	1	1	1	1.00

The parents of the missing person in Case 2 may be said to be about 44 times more likely to pass the stain phenotype than would be a randomly selected couple. Similarly, from Table 2.9, this evidence also provides a great increase in support for the hypothesis that the blood stains came from a child of the case couple.

6.8 PATERNITY—COMBINATION OF LIKELIHOOD RATIOS

In paternity testing, the likelihood ratio is used to compare two probabilities, as in the criminal context. In this context the hypotheses to be compared are:

- F: the alleged father is the true father,
- $\bar{F}$: the alleged father is not the true father.

The probability that the alleged father would pass the child's non-maternal genes is compared to the probability that the genes would be passed randomly. Thus the value V of the evidence is

$$V = \frac{\text{Probability the alleged father would pass the genes}}{\text{Probability the genes would be passed by a random male}}.$$

Table 6.10   Two pieces of evidence on phenotypes

Evidence	System	Phenotypes		
		Child	Mother	Alleged Father
E_1	ABO	O	O	O
E_2	Gc	1–2	1–1	2–2

This may be expressed verbally as 'the alleged father is V times more likely than a randomly selected man to pass this set of genes'.

Notice the difference between paternity testing and a missing persons case. In paternity testing, the relationship between mother and child is known. The question is whether a particular male is a possible biological father. In missing person/blood stain cases the parents are known. The question is whether the stain could have come from one of their offspring.

Berry and Geisser (1986) give an example of the use of the likelihood ratio in paternity problems. See also Berry (1991b) in Aitken and Stoney (1991). The ratio in this context has also been called the *paternity index* (P.I.) by Salmon and Salmon (1980). The example discussed here is the example used by Berry and Geisser (1986) and is another example of the use of the likelihood ratio to include more than one piece of evidence through consideration of more than one genetic marker.

Consider two pieces of evidence E_1 and E_2 where E_1 and E_2 are the phenotypes of child, mother and alleged father under the *ABO* and *Gc* systems, respectively, as shown in Table 6.10.

The likelihood ratio, or paternity index *P.I.*, for E_i, $i = 1, 2$, is

$$\frac{Pr(E_i|F)}{Pr(E_i|\bar{F})}.$$

Since phenotype O crossed with phenotype O will always produce O, $Pr(E_1|F) = 1$. If $\bar{F}$ holds, the child will be O if and only if an O allele is selected from the father, given the mother is O. Thus, $Pr(E_1|\bar{F})$ is the frequency of allele O, γ_O, say, in the *ABO* system of the population. Then

$$\frac{Pr(E_1|F)}{Pr(E_1|\bar{F})} = \frac{1}{\gamma_O}.$$

Consider E_2. Since 1–1 crossed with 2–2 will always produce 1–2, $Pr(E_2|F) = 1$. If $\bar{F}$ holds, the child will be 1–2 if and only if a 2 allele is selected from the father. Thus $Pr(E_2|\bar{F})$ is the frequency of allele 2 in the Gc system, γ_2 say, (Grunbaum *et al.*,

1978) and

$$\frac{Pr(E_2|F)}{Pr(E_2|\bar{F})} = \frac{1}{\gamma_2}.$$

Under an assumption of independence, the likelihood ratio for the combination (E_1, E_2) of evidence is

$$\frac{Pr(E_1, E_2|F)}{Pr(E_1, E_2|\bar{F})} = \frac{Pr(E_1|F)}{Pr(E_1|\bar{F})} \times \frac{Pr(E_2|F)}{Pr(E_2|\bar{F})}$$

$$= \frac{1}{\gamma_0} \times \frac{1}{\gamma_2}.$$

(See Section 5.1.3) From Berry and Geisser (1986), $\gamma_0 = 0.692$, $\gamma_2 = 0.290$ and $1/(\gamma_0\gamma_2) \simeq 5$. Thus, the evidence of the two blood grouping systems is five times more likely if the alleged father is the true father than if he is not. From Table 2.9 this provides an increase in support for the hypothesis that the alleged father is the true father. It is this ratio which is known as the paternity index.

6.8.1 Likelihood of paternity

In the context of paternity, it is appropriate to make a digression from consideration solely of the likelihood ratio and to consider the probability that the alleged father is the true father; i.e., the probability that F is true. This is known as the *likelihood of paternity*.

First, consider E_1. The odds in favour of F, given E_1 may be written, using the odds form of Bayes' Theorem (2.4), as

$$\frac{Pr(F|E_1)}{Pr(\bar{F}|E_1)} = \frac{Pr(E_1|F)}{Pr(E_1|\bar{F})} \times \frac{Pr(F)}{Pr(\bar{F})}$$

and

$$Pr(\bar{F}|E_1) = 1 - Pr(F|E_1)$$

so

$$Pr(F|E_1) = \frac{Pr(E_1|F)}{Pr(E_1|\bar{F})} \times \frac{Pr(F)}{Pr(\bar{F})} \times \{1 - Pr(F|E_1)\}$$

and

$$Pr(F|E_1)\left\{1 + \frac{Pr(E_1|F)}{Pr(E_1|\bar{F})} \times \frac{Pr(F)}{Pr(\bar{F})}\right\} = \frac{Pr(E_1|F)}{Pr(E_1|\bar{F})} \times \frac{Pr(F)}{Pr(\bar{F})}.$$

so that

$$Pr(F|E_1) = \left\{ 1 + \frac{Pr(E_1|\bar{F})}{Pr(E_1|F)} \times \frac{Pr(\bar{F})}{Pr(F)} \right\}^{-1}. \tag{6.7}$$

Suppose rather unrealistically that the alleged father and only one other man (of unknown blood type) could be the true father and each is equally likely (Essen-Möller, 1938). Then

$$Pr(F) = Pr(\bar{F}) = 0.5$$

and

$$Pr(F|E_1) = 1/(1 + 0.692) = 0.591.$$

Now include E_2. The posterior odds $Pr(F|E_1)/Pr(\bar{F}|E_1)$ in favour of F, given E_1, now replace the prior odds $Pr(F)/Pr(\bar{F})$ (see Section 5.1.3) and the posterior probability for F, given E_1 and E_2, is given by

$$Pr(F|E_1, E_2) = \left\{ 1 + \frac{Pr(\bar{F}|E_1)}{Pr(F|E_1)} \times \frac{Pr(E_2|\bar{F})}{Pr(E_2|F)} \right\}^{-1} \tag{6.8}$$

$$= \left(1 + \frac{0.409}{0.591} \times \frac{0.290}{1} \right)^{-1}$$

$$= 0.833,$$

where the assumption of the independence of E_1 and E_2 has been made. The alleged father's likelihood of paternity was initially 0.5. After presentation of the *ABO* evidence (E_1) it became 0.591. After the presentation of the *Gc* evidence (E_2) it became 0.833. Notice that the assumption $Pr(F) = Pr(\bar{F}) = 0.5$ is unrealistic and can lead to breaches of the laws of probability. If there were two alleged fathers, both of type $(O, 2-2)$ then both would have a posterior probability of 0.833 of being the true father. The probability of one or other being the true father would then be the sum of these two probabilities, i.e., 1.666. However, this is greater than 1 and breaches the first law of probability (1.4).

The effect on the posterior probability of altering the prior probability can be determined from (6.7) and (6.8). Some sample results are given in Table 6.11. These calculations again assume that the genetic systems are independent. This assumption is discussed by Berry and Geisser (1986) who refer to Schatkin (1984) and mention many experts who 'casually multiply probabilities for up to 100 genetic polymorphisms ... assuming independence, without saying so'. The assumption, as Berry and Geisser (1986) state, is based 'either on theoretical considerations or on experimental data. It is incumbent upon the forensic statistician to determine which, in fact, is the case and on what evidence it rests'.

Table 6.11 Posterior probabilities of paternity for various prior probabilities for evidence $E_1 = O$, $E_2 = 2$–2

$Pr(F)$	0.5	0.25	0.1	0.01
$Pr(F\|E_1)$	0.591	0.325	0.138	0.014
$Pr(F\|E_1, E_2)$	0.833	0.624	0.356	0.047

The probability $Pr(F|E_1, E_2)$ may be written as

$$\left\{ 1 + \frac{Pr(E_1|\bar{F})}{Pr(E_1|F)} \times \frac{Pr(E_2|\bar{F})}{Pr(E_2|F)} \times \frac{Pr(\bar{F})}{Pr(F)} \right\}^{-1}.$$

and if $Pr(\bar{F})$ and $Pr(F)$ are taken equal to 0.5 then

$$Pr(F|E_1, E_2) = \left\{ 1 + \frac{Pr(E_1|\bar{F})}{Pr(E_1|F)} \times \frac{Pr(E_2|\bar{F})}{Pr(E_2|F)} \right\}^{-1}.$$

In general, for n independent marker systems, giving evidence $E_1, E_2, \ldots, E_n$, with $Pr(F) = Pr(\bar{F})$,

$$Pr(F|E_1, \ldots, E_n) = \left\{ 1 + \prod_{i=1}^{n} \frac{Pr(E_i|\bar{F})}{Pr(E_i|F)} \right\}^{-1}$$

where $\prod_{i=1}^{n} Pr(E_i|\bar{F})/Pr(E_i|F)$ is the product of the reciprocals of the n likelihood ratios $Pr(E_i|F)/Pr(E_i|\bar{F})$. This expression is called the *plausibility of paternity* (Berry and Geisser, 1986). Notice that it depends on the assumption $Pr(F) = Pr(\bar{F}) = 0.5$ which is unrealistic in many cases. The assumption that $Pr(F)$ equals $Pr(\bar{F})$ may easily be dispensed with to give the following result

$$Pr(F|E_1, \ldots, E_n) = \left\{ 1 + \frac{Pr(\bar{F})}{Pr(F)} \prod_{i=1}^{n} \frac{Pr(E_i|\bar{F})}{Pr(E_i|F)} \right\}^{-1}.$$

The plausibility of paternity has also been transformed into a *likelihood of paternity* (Hummel, 1971) to provide a verbal scale, given here in Table 6.12. Notice that this verbal scale is one for probabilities. The verbal scale provided by Table 2.9, Section 2.5.3, is one for likelihood ratios.

Table 6.12 Hummel's likelihood of paternity

Plausibility of paternity	Likelihood of paternity
0.9980–0.9990	Practically proved
0.9910–0.9979	Extremely likely
0.9500–0.9909	Very likely
0.9000–0.9499	Likely
0.8000–0.8999	Undecided
Less than 0.8000	Not useful

7

Continuous Data

7.1 THE LIKELIHOOD RATIO

The previous chapter considered the evaluation of the likelihood ratio where the evidence was represented by discrete data with specific reference to genetic marker systems. The values of the evidence in various different contexts were derived. However, much evidence is of a form in which measurements may be taken and for which the data are continuous. The form of the statistic for the evaluation of the evidence under these circumstances is similar to that for discrete data. Let the evidence be denoted by E and the two competing hypotheses by C and $\bar{C}$ and background information by I, then the value V of the evidence is, formally,

$$V = \frac{Pr(E|C, I)}{Pr(E|\bar{C}, I)},$$

as before (see (2.10), Section 2.5.1 and (6.1), Section 6.1). The quantitative part of the evidence is represented by the measurements of the characteristic of interest. Let x denote the measurements on the source evidence and let y denote the measurements on the receptor object. For example, if a window is broken during the commission of a crime, the measurements on the refractive indices of m fragments of glass found at the crime scene will be denoted $x_1, \ldots, x_m$ (denoted $\mathbf{x}$). The refractive indices of n fragments of glass found on a suspect will be denoted $y_1, \ldots, y_n$ (denoted $\mathbf{y}$). The quantitative part of the evidence concerning the glass fragments in this case can be denoted by

$$E = (\mathbf{x}, \mathbf{y}).$$

In the notation of Section 1.6.1, M_c is the broken window at the crime scene, M_s is the set of glass fragments from the suspect, E_c is $\mathbf{x}$, E_s is $\mathbf{y}$, M is (M_c, M_s) and $E = (E_c, E_s) = (\mathbf{x}, \mathbf{y})$. Continuous measurements are being considered and the probabilities Pr are therefore replaced by probability density functions f (see Section 3.4.1) so that

$$V = \frac{f(\mathbf{x}, \mathbf{y}|C, I)}{f(\mathbf{x}, \mathbf{y}|\bar{C}, I)}. \tag{7.1}$$

Bayes' Theorem and the rules of conditional probability apply to probability density functions as well as to probabilities. The value, V, of the evidence (7.1) may be rewritten in the following way.

$$V = \frac{f(\mathbf{x}, \mathbf{y} | C, I)}{f(\mathbf{x}, \mathbf{y} | \bar{C}, I)}$$

$$= \frac{f(\mathbf{y} | \mathbf{x}, C, I)}{f(\mathbf{y} | \mathbf{x}, \bar{C}, I)} \times \frac{f(\mathbf{x} | C, I)}{f(\mathbf{x} | \bar{C}, I)} .$$

The measurements $\mathbf{x}$ are those on the source object. Their distribution and corresponding probability density function are independent of whether C or $\bar{C}$ is true. Thus

$$f(\mathbf{x} | C, I) = f(\mathbf{x} | \bar{C}, I)$$

and

$$V = \frac{f(\mathbf{y} | \mathbf{x}, C, I)}{f(\mathbf{y} | \mathbf{x}, \bar{C}, I)} .$$

If $\bar{C}$ is true then the measurements ($\mathbf{y}$) on the receptor object and the measurements ($\mathbf{x}$) on the source object are independent. Thus

$$f(\mathbf{y} | \mathbf{x}, \bar{C}, I) = f(\mathbf{y} | \bar{C}, I),$$

and

$$V = \frac{f(\mathbf{y} | \mathbf{x}, C, I)}{f(\mathbf{y} | \bar{C}, I)} . \tag{7.2}$$

The numerator is in the form of a distribution known as a *predictive distribution* (Aitchison and Dunsmore, 1975; Evett *et al.*, 1987) since $\mathbf{y}$ is said to be predicted from $\mathbf{x}$. The denominator is the so-called *marginal distribution* of the receptor measurements in the relevant population, the definition of which is assisted by I. This formulation of the expression for V shows that for the numerator, the distribution of the receptor measurements, conditional on the source measurements as well as I, is considered. For the denominator, the distribution of the receptor measurements is considered over the distribution of the whole of the relevant population.

The two hypotheses to be compared are:

- C: the receptor sample is from the same source as the bulk sample;
- $\bar{C}$: the receptor sample is from a different source than the bulk sample.

First, consider C. The bulk and receptor measurements are on evidence from the same source. The measurements on this source have a true mean θ, say. For example, if the measurements are of the refractive index of glass then θ denotes the mean refractive index of the window from which the fragments have been taken. For clarity, the conditioning elements C and I, the background information, will be omitted in the following argument. The predictive distribution $f(\mathbf{y}|\mathbf{x})$ may be expressed as follows:

$$
\begin{aligned}
f(\mathbf{y}|\mathbf{x}) &= \int f(\mathbf{y}|\theta)f(\theta|\mathbf{x})d\theta \\
&= \frac{\int f(\mathbf{y}|\theta)f(\mathbf{x}|\theta)f(\theta)d\theta}{f(\mathbf{x})} \\
&= \frac{\int f(\mathbf{y}|\theta)f(\mathbf{x}|\theta)f(\theta)d\theta}{\int f(\mathbf{x}|\theta)f(\theta)d\theta},
\end{aligned}
$$

the ratio of the joint distribution of $\mathbf{x}$ and $\mathbf{y}$ to the marginal distribution of $\mathbf{x}$. Both distributions are independent of θ which is integrated out. The distributions are of measurements over all windows.

For $\bar{C}$, the situation where the bulk and receptor measurements are from different sources, it is the measurements $\mathbf{y}$ on the receptor object which are the ones of interest. The probability density function for $\mathbf{y}$ is

$$
f(\mathbf{y}) = \int f(\mathbf{y}|\theta)f(\theta)d\theta.
$$

V may then be written as

$$
\frac{\int f(\mathbf{y}|\theta)f(\mathbf{x}|\theta)f(\theta)d\theta}{\int f(\mathbf{x}|\theta)f(\theta)d\theta \int f(\mathbf{y}|\theta)f(\theta)d\theta}. \tag{7.3}
$$

For those unfamiliar with these kinds of manipulations, Bayes' Theorem applied to conditional probability distributions is used to write $f(\theta|\mathbf{x})$ as $f(\mathbf{x}|\theta)f(\theta)/f(\mathbf{x})$. The law of total probability with integration replacing summation is used to write $f(\mathbf{x})$ as $\int f(\mathbf{x}|\theta)f(\theta)d\theta$.

7.2   NORMAL DISTRIBUTION FOR BETWEEN SOURCE DATA

The approach to evidence evaluation described above was first proposed by Lindley (1977a) in the context of a problem involving the measurements of the refractive index of glass. These measurements may be made on fragments of glass at the scene of a crime and on fragments of window glass found on a suspect's clothing (see Example 1.2). These measurements are subject to error and it is this which is incorporated into V.

7.2.1   Sources of variation

Notice that there are often two sources of variation to be considered in the measurements. There is variation within a particular source and there is variation between sources.

For example, consider evidence of fragments of glass from a broken window from which refractive index (r.i.) measurements have been made. There is variation in the r.i. measurements amongst the different fragments of glass. These different measurements may be thought of as a sample from the population corresponding to all possible r.i. measurements from that particular window. The population has a mean, θ say, and a variance σ^2. The measurements of r.i. of fragments from that window are assumed to be Normally distributed with mean

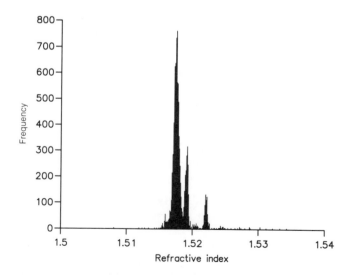

Figure 7.1   Refractive index measurements from 2265 fragments of float glass from buildings (Lambert and Evett, 1984).

θ and variance σ^2. Secondly, there is variation in the r.i. mean θ between different windows. The mean θ has a probability distribution with its own mean μ, say, and variance τ^2. Typically τ^2 will be much greater than σ^2. Initially it will be assumed that θ has a Normal distribution also. However, a look at Figure 7.1 which is a histogram of r.i. measurements from 2269 examples of float glass from buildings given in Table 7.5 (Lambert and Evett, 1984) shows that this is not a particularly realistic assumption. A more realistic approach will be described in Section 7.4

Similar considerations apply for other types of evidence. For measurements on the medullary widths of cat hairs, for example, there will be variation amongst hairs from the same cat and amongst hairs from different cats. For human head hairs there will be variations in characteristics between hairs from the same head (and also within the same hair) and between hairs from different heads.

Thus, when considering the assessment of continuous data at least two sources of variability have to be considered: the variability within the source (e.g. window) from which the measurements were made and the variability between the different possible sources (e.g., windows).

7.2.2 Derivation of the marginal distribution

Let x be a measurement from a particular bulk fragment. Let the mean of measurements from the source of this fragment be θ_1. Let y be one measurement from a particular receptor fragment. Let the mean of measurements from the source of this fragment be θ_2. The variance of measurements within a source is assumed constant amongst sources and is denoted σ^2. The dependence of the distribution of these measurements on the source from which they come can be made notationally explicit. The distributions of X and Y, given θ_1, θ_2 and σ^2 are

$$(X|\theta_1, \ \sigma^2) \sim N(\theta_1, \ \sigma^2),$$
$$(Y|\theta_2, \ \sigma^2) \sim N(\theta_2, \ \sigma^2),$$

where the dependence on θ_1 or θ_2 and σ^2 is made explicit. Notice, also, that variation in X is modelled. Contrast this approach with the coincidence probability approach of Section 4.6 in which the mean of measurements on the bulk fragments was taken as fixed. The conditioning on $\overline{C}$ is implicit. The means θ_1 and θ_2 of these distributions may themselves be thought of as observations from another distribution (that of variation between sources) which for the present is taken to be Normal, with mean μ and variance τ^2. Thus, θ_1 and θ_2 have the same probability density function and

$$(\theta|\mu, \tau^2) \sim N(\mu, \tau^2).$$

The distributions of X and of Y, independent of θ, can be determined by taking the so-called *convolutions* of x and of y with θ to give

$$f(x|\mu, \sigma^2, \tau^2) = \int f(x|\theta, \sigma^2)f(\theta|\mu, \tau^2)d\theta$$

$$= \int \frac{1}{2\pi\sigma^2\tau^2} \exp\left\{-\frac{1}{2\sigma^2}(x-\theta)^2\right\} \exp\left\{-\frac{1}{2\tau^2}(\theta-\mu)^2\right\} d\theta$$

$$= \frac{1}{\sqrt{\{2\pi(\sigma^2+\tau^2)\}}} \exp\left\{-\frac{1}{2(\sigma^2+\tau^2)}(x-\mu)^2\right\},$$

using the result that

$$\frac{1}{2\sigma^2}(x-\theta)^2 + \frac{1}{2\tau^2}(\theta-\mu)^2 = \frac{(\theta-\mu_1)^2}{\tau_1^2} + \frac{(x-\mu)^2}{\sigma^2+\tau^2}$$

where

$$\mu_1 = \frac{\sigma^2\mu + \tau^2 x}{\sigma^2+\tau^2},$$

$$\tau_1^2 = \frac{\sigma^2\tau^2}{\sigma^2+\tau^2}.$$

Similarly

$$f(y|\mu, \sigma^2, \tau^2) = \frac{1}{\sqrt{\{2\pi(\sigma^2+\tau^2)\}}} \exp\left\{-\frac{1}{2(\sigma^2+\tau^2)}(y-\mu)^2\right\}.$$

Notice that τ^2 has been omitted from the distributions of x and y, given θ_1, θ_2 and σ^2. This is because the distributions of x and y, given these parameters, are independent of τ^2. Similarly, the distribution of θ, given μ and τ^2, is independent of σ^2.

The effect of the two sources of variability is that the mean of the r.i. measurements is the overall mean μ and the variance is the sum of the two component variances σ^2 and τ^2. The distribution remains Normal:

$$(X|\mu, \sigma^2, \tau^2) \sim N(\mu, \sigma^2+\tau^2),$$

$$(Y|\mu, \sigma^2, \tau^2) \sim N(\mu, \sigma^2+\tau^2).$$

7.2.3 Approximate derivation of the likelihood ratio

Consider an application to a broken window as in Example 1.2. A crime is committed in which a window is broken. A suspect is apprehended soon afterwards and a fragment of glass is found on his clothing. Its refractive index is y. A sample of m fragments is taken from the broken window at the scene of the crime and their refractive index measurements are $\mathbf{x} = (x_1, \ldots, x_m)$ with mean $\bar{x}$.

The two hypotheses to be compared are:

- C: the receptor fragment is from the crime scene window;
- $\bar{C}$: the receptor fragment is not from the crime scene window.

An approximate derivation of the likelihood ratio may be obtained by replacing θ by $\bar{x}$ in the distribution of y so that $f(y|\theta, \sigma^2)$ becomes $f(y|\bar{x}, \sigma^2)$. This is only an approximate distributional result. A more accurate result is given later to account for the sampling variability of $\bar{x}$. For the present, an approximate result for the numerator is that

$$(Y|\bar{x}, \sigma^2, C, I) \sim N(\bar{x}, \sigma^2).$$

Also, an approximate result for the denominator is that

$$(Y|\mu, \sigma^2, \tau^2, \bar{C}, I) \sim N(\mu, \tau^2 + \sigma^2).$$

For τ^2 much greater than σ^2, assume also that $\tau^2 + \sigma^2$ can be approximated by τ^2. The likelihood ratio is than

$$V = \left[\frac{1}{\sigma\sqrt{(2\pi)}} \exp\left\{ -\frac{(y - \bar{x})^2}{2\sigma^2} \right\} \right] \Big/ \left[\frac{1}{\tau\sqrt{(2\pi)}} \exp\left\{ -\frac{(y - \mu)^2}{2\tau^2} \right\} \right]$$

$$= \frac{\tau}{\sigma} \exp\left\{ \frac{(y - \mu)^2}{2\tau^2} - \frac{(y - \bar{x})^2}{2\sigma^2} \right\},$$

(Evett, 1986).

Note that this likelihood ratio depends on an assumption that fragments from a single source are found on the suspect and that these have come from the crime scene.

This result has some intuitively attractive features. The likelihood ratio is larger for values of y which are further from μ and are therefore assumed to be more rare; i.e., the rarer the value of the refractive index of the recovered fragment the larger the likelihood ratio. Also, the larger the value of $|y - \bar{x}|$ the smaller the value of the likelihood ratio; i.e., the further the value of the refractive index of the receptor glass fragment is from the mean of the values of the refractive indices of the source fragments, the smaller the likelihood ratio.

Table 7.1 Likelihood ratio values for varying values of $(y - \bar{x})/\sigma$ and $(y - \mu)/\tau$

| $(y - \bar{x})/\sigma$ | $(y - \mu)/\tau$ | | |
	0	1	2
0	100	165	739
1	61	100	448
2	14	22	100
3	1	2	8

Values for τ equal to 4×10^{-3} and for σ equal to 4×10^{-5} are given by Evett (1986). Values of V for various values of $(y - \mu)/\tau$ and of $(y - \bar{x})/\sigma$, the standardised distances of y from the overall mean and the source mean, are given in Table 7.1. Note that $\tau/\sigma = 100$, giving ample justification for the approximation τ^2 to the variance of y, given $\bar{C}$, above.

Consideration of the probabilities of transfer of fragments (Section 5.3.3), both from the crime scene and by innocent means from elsewhere (Evett, 1986) is left till later (Section 7.5).

7.2.4 Lindley's approach

A more detailed analysis was provided by Lindley (1977a). Assume, as before, that the measurements are distributed about the true unknown value, θ, of the refractive index with a Normal distribution and a known constant variance σ^2 and that the hypotheses C and $\bar{C}$ to be compared are as in Section 7.2.3. If m measurements are made at the scene (source measurements, $x_1, \ldots, x_m$) then it is sufficient to consider their mean, $\bar{x} = \sum_{i=1}^{m} x_i/m$. Conditional on θ_1, the mean of the refractive index of the crime window, the mean $\bar{x}$ is Normally distributed about θ_1 with variance σ^2/m. Let $\bar{y}$ denote the mean of n similar measurements (receptor measurements, $y_1, \ldots, y_n$) made on material found on the suspect; conditional on θ_2, $\bar{y}$ is Normally distributed about θ_2 with variance σ^2/n. In the case C where the source and receptor measurements come from the same source, $\theta_1 = \theta_2$. Otherwise, in the case $\bar{C}$, $\theta_1 \neq \theta_2$.

The distribution of the true values θ has also to be considered. There is considerable evidence about the distribution of refractive indices; see, for example, Lambert and Evett (1984). First, assume as before that the true values θ are Normally distributed about a mean μ with variance τ^2, both of which are assumed known. Typically τ will be larger, sometimes much larger, than σ (see above, where $\tau/\sigma = 100$). This assumption of Normality is not a realistic one in this context where the distribution has a pronounced peak and a long tail to the right; see Figure 7.1. However, the use of the Normality assumption enables analytic results to be obtained as an illustration of the general application of the method. The marginal distributions of $\bar{X}$ and $\bar{Y}$ are $N(\mu, \tau + \sigma^2/m)$ and $N(\mu, \tau + \sigma^2/n)$.

A brief derivation of V is given in an Appendix to this chapter (Section 7.6.1) and follows the arguments of Lindley (1977a). From Section 7.6.1, if the number of control measurements equals the number of recovered measurements, $m = n$, $z = w = \frac{1}{2}(\bar{x} + \bar{y})$ and $\sigma_1^2 = \sigma_2^2$. Then

$$V \simeq \frac{m^{1/2}\tau}{2^{1/2}\sigma} \times \exp\left\{-\frac{m(\bar{x} - \bar{y})^2}{4\sigma^2}\right\} \times \exp\left\{\frac{(z - \mu)^2}{2\tau^2}\right\}. \qquad (7.4)$$

Note again that this result assumes implicitly that the fragments are from a single source. Denote this assumption by S. Then the above result is the ratio of the probability density functions $f(\bar{x}, \bar{y} | C, S)/f(\bar{x}, \bar{y} | \bar{C}, S)$. A result including S as one of the uncertain elements and deriving an expression for $f(\bar{x}, \bar{y}, S | C)/f(\bar{x}, \bar{y}, S | \bar{C})$ was given by Grove (1980). Let T denote the event that fragments were transferred from the broken window to the suspect and persisted there until discovery by the police. Let A be the event that the suspect came into contact with glass from some other source. Assume that $Pr(A|C) = Pr(A|\bar{C}) = p_A$, that $Pr(T|C) = p_T$, that A and T are independent given C and that $Pr(T|\bar{C}) \simeq 0$, then Grove (1980) shows that

$$V = \frac{f(\bar{x}, \bar{y}, S | C)}{f(\bar{x}, \bar{y}, S | \bar{C})}$$

$$= 1 + p_T\left\{(p_A^{-1} - 1)\frac{f(\bar{x}, \bar{y}|C)}{f(\bar{x}, \bar{y}|\bar{C})} - 1\right\}$$

$$= (1 - p_T) + \frac{p_T(1 - p_A)}{p_A} \times \frac{f(\bar{x}, \bar{y}|C)}{f(\bar{x}, \bar{y}|\bar{C})}, \qquad (7.5)$$

where $f(\bar{x}, \bar{y}|C)/f(\bar{x}, \bar{y}|\bar{C})$ is Lindley's (1977a) ratio. The value derived by Grove (1980) takes account of transfer and persistence in a way already derived for discrete data (see Section 5.3.3). Another derivation for continuous data is given by (7.10) in Section 7.5.2.

7.2.5 Interpretation of result

The interpretation of (7.4) is now considered in the particular case $m = n = 1$. V consists of two factors which depend on the measurements. The first is $\exp\{-(\bar{x} - \bar{y})^2/4\sigma^2\}$. This compares the absolute difference $|\bar{x} - \bar{y}|$ of the control and recovered measurements with their standard deviation $\sigma\sqrt{2}$ on the hypothesis $(\theta_1 = \theta_2)$ that they come from the same source. Let $|\bar{x} - \bar{y}|/\sigma\sqrt{2} = \lambda$. Then the value of the first factor is $\exp(-\lambda^2/2)$. A large value of λ favours the hypothesis that the two fragments come from different sources. This factor has an effect like that of a significance test of a null hypothesis of identity ($\theta_1 = \theta_2$).

The second factor, $\exp\{(z-\mu)^2/2\tau^2\}$, with $z = \frac{1}{2}(\bar{x} + \bar{y})$ measures the typicality of the two measurements. This factor takes its smallest value, 1, when $z = \mu$ and increases as $|z - \mu|$ increases relative to its standard deviation. Thus the more unusual the glass (i.e., the larger the value of $|z - \mu|$), the greater the value of V and the stronger the inference in favour of a common source for the two measurements. Consider the comment by Parker and Holford (1968) in Section 4.7. The first factor considers similarity. The second factor considers typicality. The assessment of similarity is not by-passed.

7.2.6   Examples

Assume that $m = n = 1$, $\tau/\sigma = 100$, $|\bar{x} - \bar{y}|/\sigma\sqrt{2} = 2$, $z = \mu$. Then

$$V = \frac{100\,e^{-2}}{\sqrt{2}} = 9.57.$$

The odds in favour of a common source are increased by a factor of almost 10. (Note that $\bar{x}$ and $\bar{y}$ are 2 standard deviations apart. A conventional significance test, Section 4.5.1, would have rejected the hypothesis of a common source at the 5% level of significance.)

The values of (7.4) for $\tau/\sigma = 100$ as a function of λ and $\delta = |z - \mu|/\tau$, the deviation of the mean of the two measurements from μ, standardised on the assumption that the hypothesis of a common source is true, are given in Table 7.2.

Consider the more general formula for V, as given in (7.16) in Section 7.6, namely

$$V \simeq \frac{\tau}{a\sigma} \times \exp\left\{-\frac{(\bar{x}-\bar{y})^2}{2a^2\sigma^2}\right\} \times \exp\left\{-\frac{(w-\mu)^2}{2\tau^2} + \frac{(z-\mu)^2}{\tau^2}\right\}. \qquad (7.6)$$

Table 7.2   Value of $\tau(2^{1/2}\sigma)^{-1}\exp(-\frac{1}{2}\lambda^2 + \frac{1}{2}\delta^2)$ (7.4) as a function of $\lambda = |\bar{x} - \bar{y}|/(2^{1/2}\sigma)$ and $\delta = |z - \mu|/\tau$ for $\tau/\sigma = 100$, $m \simeq 1$

| δ | λ | | | | |
	0	1.0	2.0	4.0	6.0
0	70.7	42.9	9.57	0.024	1.08×10^{-6}
1.0	117	70.7	15.8	0.039	1.78×10^{-6}
2.0	522	317	70.7	0.175	7.94×10^{-6}
3.0	6370	3860	861	2.14	9.71×10^{-5}

The following information is needed in order that V may be evaluated:

- The number of source measurements (m).
- The mean of the source measurements ($\bar{x}$).
- The number of receptor measurements (n).
- The mean of the receptor measurements ($\bar{y}$).
- The variance (assumed known) of the measurements on the source and receptor samples (σ^2).
- The overall mean (assumed known) of the refractive indices (μ).
- The overall variance (assumed known) of the refractive indices (τ^2).

The following values may be derived from the above:

- $z = (\bar{x} + \bar{y})/2$.
- $w = (m\bar{x} + n\bar{y})/(m + n)$.
- $a^2 = 1/m + 1/n$.

Consider the following numerical example using data from Evett (1977) and Lindley (1977a) where $\bar{x} = 1.518\,458$, $m = 10$; $\bar{y} = 1.518\,472$, $n = 5$; $\sigma = 0.000\,04$; $\tau = 0.004$. The overall mean μ is taken to be 1.5182 and has been derived from the 2269 measurements for building float glass published by Lambert and Evett (1984); see Figure 7.1. With these figures, $a^2 = 0.3$, $w = 1.518\,463$, $z = 1.518\,465$, and

$$\tau/a\sigma = 182.5742,$$
$$(\bar{x} - \bar{y})^2/2a^2\sigma^2 = 0.2042,$$
$$(w - \mu)^2/2\tau^2 = 0.002\,16,$$
$$(z - \mu)^2/\tau^2 = 0.004\,39,$$
$$V = 149.19.$$

The odds in favour of the suspect being at the crime scene are thus increased by a factor of 150.

7.3 ESTIMATION OF A PROBABILITY DENSITY FUNCTION

The estimation of a population mean (μ) and variance (σ^2) by a sample mean ($\bar{x}$) and variance (s^2) of data sampled from the population is a common idea. What is not so common is the idea that a probability density function itself may be estimated from data taken from the population. That such a procedure is necessary becomes apparent when it is realised that not all data have a distribution which is readily modelled by a standard distribution. In particular, not all

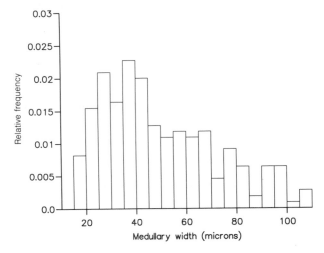

Figure 7.2   Medullary widths (in microns) of 220 cat hairs (Peabody *et al.*, 1983).

data are unimodal, symmetric and bell-shaped and may not be modelled by a Normal distribution. The histogram of the refractive index of glass fragments (Figure 7.1) and a histogram of the medullary width of cat hairs (Figure 7.2 from data in Table 7.3) both illustrate this.

Estimation of a probability density function is not too difficult so long as the distribution is fairly smooth. A procedure known as *kernel density estimation* is used; see Silverman (1986) for technical details. For early applications to forensic science, see Aitken and MacDonald (1979) for an application with discrete data to forensic odontology and Aitken (1986) for an application to the discrimination between cat and dog hairs. An example is given here of the application of the technique to the distribution of the medullary width of cat hairs.

Consider data on the medullary widths (in microns) of 220 cat hairs (Peabody *et al.*, 1983). A version of these modified to make the analysis easier is given in Table 7.3 and a histogram to illustrate the distribution is shown in Figure 7.2 from which it can be seen that the data are positively skewed and perhaps not unimodal. The histogram has been constructed from the full data set by selecting intervals of fixed width and fixed boundary points, namely (15.0–20.0, 20.01–25.00, ..., 105.01–110.00) microns. Individual observations are then allocated to the appropriate interval and a frequency count obtained. Each interval is five units (microns) wide and there are 220 observations. If each observation is allocated unit height, the total area encompassed by the histogram is 5×220 equals 1100 units. Thus if the height of each bar of the histogram is reduced by a factor of 1100, the area under the new diagram is 1. This new histogram may be

Table 7.3 Medullary widths in microns of 220 cat hairs (Peabody *et al.*, 1983)

17.767	28.600	39.433	52.233	68.467
18.633	28.600	39.867	52.867	69.333
19.067	29.033	39.867	53.300	71.067
19.067	29.033	39.867	53.300	71.500
19.067	29.467	40.300	53.733	71.667
19.133	30.333	40.300	53.733	73.233
19.300	30.767	40.733	54.167	74.533
19.933	31.200	41.167	54.600	75.400
19.933	31.200	41.167	55.033	76.267
20.367	31.300	41.600	55.467	76.267
20.367	31.633	41.600	55.900	77.133
20.367	31.633	42.033	56.767	77.367
20.600	31.807	42.467	57.200	78.000
20.800	32.000	42.467	57.200	78.000
20.800	32.067	42.467	57.200	79.500
21.233	32.067	42.467	57.633	79.733
21.233	32.500	42.467	58.067	80.167
21.400	32.500	42.467	58.067	80.167
22.533	33.800	42.900	58.500	80.167
22.967	33.800	42.900	58.933	81.467
22.967	34.233	42.900	58.933	81.467
23.400	34.667	42.900	60.233	81.900
23.833	34.667	43.333	60.533	82.767
23.833	35.533	44.200	60.667	84.067
24.267	35.533	44.200	60.667	87.100
24.700	35.533	44.300	60.667	87.967
25.133	35.533	45.067	61.100	90.133
25.133	35.533	45.933	61.967	90.267
25.133	36.400	45.933	62.400	91.867
25.133	36.400	45.933	62.400	91.867
25.300	37.267	46.150	63.000	92.733
26.000	37.267	46.583	63.267	93.167
26.000	37.267	46.800	63.700	93.600
26.233	38.567	46.800	65.433	95.333
26.433	38.567	47.167	65.867	96.267
26.433	38.567	48.100	66.300	97.067
26.867	39.000	48.317	66.733	97.500
26.867	39.000	48.967	66.733	97.500
27.133	39.000	48.967	66.733	97.933
27.733	39.000	49.400	67.167	99.667
27.733	39.000	50.267	67.600	100.100
27.733	39.433	51.567	67.600	106.600
28.167	39.433	51.567	68.033	106.600
28.167	39.433	52.000	68.033	107.467

considered a very naive probability function (with steps at the boundary points of the bars of the histogram).

The method of kernel density estimation may be considered as a development of the histogram. Consider the histogram to be constructed with rectangular blocks, each block corresponding to one observation. The block is positioned according to the interval in which the observation lies. The method of kernel density estimation used here replaces the rectangular block by a Normal probability density curve, known in this context as the *kernel function*. The curve is positioned by centring it over the observation to which it relates. The estimate of the probability density curve is then obtained by adding the individual curves together over all the observations in the data set and then dividing this sum by the number of observations. Since each component of the sum is a probability density function, each component has area 1. Thus, the sum of the functions divided by the number of observations also has area 1 and is a probability density function.

In the construction of a histogram a decision has to be made initially as to the width of the intervals. If these are wide, the histogram is very uninformative regarding the underlying distribution. If these are narrow, there is too much detail and general features of the distribution are lost. Similarly, in kernel density estimation, the spread of the Normal density curves has to be determined. The spread of the curves is represented by the variance. If the variance is chosen to be large, the resultant estimated curve is very smooth. If the variance is chosen to be small the resultant curve is very spikey (see Figure 7.3).

Mathematically, the kernel density estimate of an underlying probability density function can be constructed as follows. The discussion is in the context of estimating the distribution of the medullary widths of cat hairs. There is variation in the medullary width both within hairs from an individual cat and between different cats. Denote the measurement of the mean medullary width of hairs from a particular cat by θ. The corresponding probability density function $f(\theta)$ is to be estimated. A data set $D = \{z_1, \ldots, z_k\}$ is available to enable this to be done. The variance of the width of hairs from different cats is estimated by

$$s^2 = \sum_{i=1}^{k} (z_i - \bar{z})^2/(k-1)$$

where $\bar{z}$ is the sample mean. This variance is a mixture of the variances measuring the variability of the medullary width between and within cats and will be used as an approximation to the variance of the medullary width between cats. The sample standard deviation s is then multiplied by a parameter, known as the *smoothing parameter*, denoted here by λ, which determines the smoothness of the density estimate. The kernel density function $K(\theta|z_i, \lambda)$ for point z_i is then taken to be a Normal distribution with mean z_i and variance $\lambda^2 s^2$,

$$K(\theta|z_i, \lambda) = \frac{1}{\lambda s \sqrt{2\pi}} \exp\left\{-\frac{(\theta - z_i)^2}{2\lambda^2 s^2}\right\}.$$

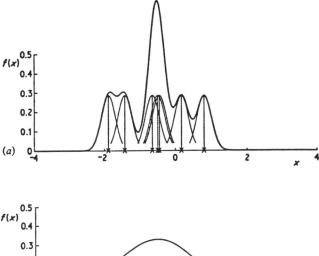

Figure 7.3 Kernel estimates showing individual kernels. Smoothing parameter values (a) 0.2, (b) 0.8 (Silverman, 1986) (Reproduced with permission of Chapman and Hall.)

The estimate $\hat{f}(\theta|D, \lambda)$ of the probability density function is then given by

$$\hat{f}(\theta|D, \lambda) = \frac{1}{k} \sum_{i=1}^{k} K(\theta|z_i, \lambda). \qquad (7.7)$$

Notice here that there is an implicit assumption that a suitable data set D exists. Also, it is assumed that it is a data set from a relevant population. This latter comment is of particular relevance when considering DNA profiling (see Chapter 8) where there is much debate as to the choice of the relevant population in a particular case.

The smoothing parameter λ has to be chosen. Mathematical procedures exist which enable an automatic choice to be made. For example, a so-called *pseudo-maximum likelihood* procedure (Habbema *et al.*, 1974) was used to determine the value of λ (0.09) used in Figure 7.4. A value of λ equal to 0.50 was used to produce the curve in Figure 7.5 from which the effect that a larger value of λ produces a smoother curve is illustrated.

The choice of λ has to be made bearing in mind that the aim of the analysis is to provide a value V for the evidence in a particular case, as represented by the likelihood ratio. Using the kernel density estimation procedure an expression for

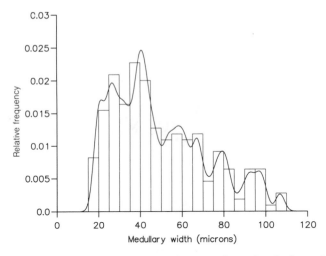

Figure 7.4   Medullary widths (in microns) of 220 cat hairs (Peabody *et al.*, 1983) and associated kernel density estimate with smoothing parameter value 0.09.

V is derived; see (7.9). An investigation of the variation in V as λ varies is worthwhile. If V does not vary greatly as λ varies, then a precise value for λ is not necessary. For example, it is feasible to choose λ subjectively by comparing the density estimate curve $\hat{f}$ obtained for various values of λ with the histogram of the data. The value which provides the best visual fit can then be chosen. Alternatively, from a scientist's personal experience of the distribution of the measurements on the characteristic of interest, it may be thought that certain possible values are not fully represented in the data set D available for estimation. In such a situation a larger value of λ may be chosen in order to provide a smoother curve, more representative of the scientist's experience. The subjective comparison of several plots of the data, produced by smoothing by different amounts, may well help to give a greater understanding of the data than the consideration of one curve, produced by an automatic method.

The choice of λ is also sensitive to outlying observations. The original cat hair data included one hair with a medullary width over 139 microns, the next largest being under 108 microns. The value of λ chosen by the automatic pseudo-maximum likelihood procedure was 0.35, a value which produced a very different estimate of the probability density function from that produced by the value of λ of 0.09 when the data set was modified as has been done by replacing the value of 139 microns by a value of 63 microns. The choice of λ is also difficult if the data are presented in grouped form as is the case with the glass data (Table 7.5). In this case, the value of λ was chosen subjectively, see Figures 7.6 and 7.7 with values of λ of 0.025 and 0.25.

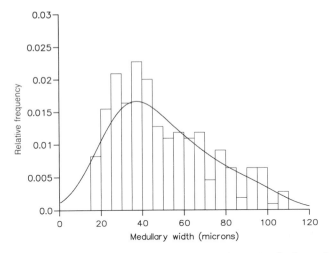

Figure 7.5 Medullary widths (in microns) of 220 cat hairs (Peabody *et al.*, 1983) and associated kernel density estimate with smoothing parameter value 0.50.

7.4 KERNEL DENSITY ESTIMATION FOR BETWEEN SOURCE DATA

If the assumption of a Normal distribution for θ is thought unrealistic the argument may be modified for a general distribution for θ using kernel density estimation as described by Chan and Aitken (1989) for cat hairs and by Berry (1991a) and Berry *et al.* (1992) for DNA profiling.

An application to the evaluation of fibre evidence in which the marginal distribution of the receptor measurements y themselves was estimated by a kernel density function, was given by Evett *et al.* (1987) and a rather more elaborate treatment was given by Wakefield *et al.* (1991). However, this was a bivariate case involving colour measurements, details of which are not discussed in this book. The method described here is applicable to situations in which the data are univariate and for which there are two components of variability, that within a particular source (e.g., window or cat) and that between different sources (e.g. windows or cats).

Consider the numerator in the original expression (7.3) for V, namely,

$$\int f(\mathbf{y}\,|\,\theta)f(\mathbf{x}\,|\,\theta)f(\theta)\mathrm{d}\theta.$$

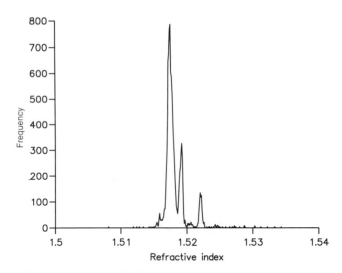

Figure 7.6   Kernel density estimate with smoothing parameter value 0.025 of refractive index measurements from 2,265 fragments of float glass from buildings (Lambert and Evett, 1984).

Given the value for θ, the distribution of $(\bar{x} - \bar{y})$ is $N(0, a^2\sigma^2)$ and the distribution of W, given θ, is $N(\theta, \sigma^2/(m + n))$. If a change in the numerator is made from $(\bar{x}, \bar{y})$ to $(\bar{x} - \bar{y}, w)$ then V may be written as

$$\frac{\dfrac{1}{a\sigma}\exp\left\{-\dfrac{(\bar{x}-\bar{y})^2}{2a^2\sigma^2}\right\}\displaystyle\int\dfrac{(m+n)^{1/2}}{\sigma}\exp\left\{-\dfrac{(w-\theta)^2(m+n)}{2\sigma^2}\right\}f(\theta)d\theta}{\displaystyle\int\dfrac{\sqrt{m}}{\sigma}\exp\left\{-\dfrac{(\bar{x}-\theta)^2m}{2\sigma^2}\right\}f(\theta)d\theta\int\dfrac{\sqrt{n}}{\sigma}\exp\left\{-\dfrac{(\bar{y}-\theta)^2n}{2\sigma^2}\right\}f(\theta)d\theta} \tag{7.8}$$

(Lindley, 1977a). The probability density function, $f(\theta)$, for θ was previously assumed to be Normal. If this is thought to be unrealistic the probability density function may be estimated by kernel density estimation.

The expression for V in (7.8) may be evaluated when $f(\theta)$ is replaced by the expression in (7.7). Some straightforward, but tedious, mathematics gives the result that

$$V = \frac{K\exp\left\{-\dfrac{(\bar{x}-\bar{y})^2}{2a^2\sigma^2}\right\}\displaystyle\sum_{i=1}^{k}\exp\left\{-\dfrac{(m+n)(w-z_i)^2}{2[\sigma^2+(m+n)s^2\lambda^2]}\right\}}{\displaystyle\sum_{i=1}^{k}\exp\left\{-\dfrac{m(\bar{x}-z_i)^2}{2(\sigma^2+ms^2\lambda^2)}\right\}\sum_{i=1}^{k}\exp\left\{-\dfrac{n(\bar{y}-z_i)^2}{2(\sigma^2+ns^2\lambda^2)}\right\}} \tag{7.9}$$

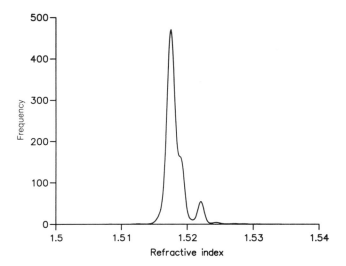

Figure 7.7 Kernel density estimate with smoothing parameter value 0.25 of refractive index measurements from 2265 fragments of float glass from buildings (Lambert and Evett, 1984).

where

$$K = \frac{k\sqrt{(m+n)}\sqrt{(\sigma^2 + ms^2\lambda^2)}\sqrt{(\sigma^2 + ns^2\lambda^2)}}{a\sigma\sqrt{(mn)}\sqrt{\{\sigma^2 + (m+n)s^2\lambda^2\}}}.$$

There are four factors, explicitly dependent on the data, in the expression for V which contribute to its overall value.

(a) $\exp\{-(\bar{x} - \bar{y})^2/(2a^2\sigma^2)\}$;

(b) $\displaystyle\sum_{i=1}^{k} \exp\{-(m+n)(w - z_i)^2/2[\sigma^2 + (m+n)s^2\lambda^2]\}$;

(c) $\displaystyle\sum_{i=1}^{k} \exp\{-m(\bar{x} - z_i)^2/2(\sigma^2 + ms^2\lambda^2)\}$;

(d) $\displaystyle\sum_{i=1}^{k} \exp\{-n(\bar{y} - z_i)^2/2(\sigma^2 + ns^2\lambda^2)\}$.

The first factor, (a), accounts for the difference between the source and receptor evidence. A large difference leads to a smaller value of V, a small difference to a larger value of V.

The second factor, (b), accounts for the location of the combined evidence in the overall distribution from the relevant population. If it is far from the centre of this distribution then V will be smaller than if it were close. This provides a measure of the rarity of the combined evidence.

The third and fourth factors, (c) and (d), account for the rarity or otherwise of the source and receptor evidence, separately. The further these are from the centre of the overall distribution the smaller the corresponding factor and the larger the value of V.

Notice, also, the difference between σ^2 which measures the variance within a particular source (e.g. window or cat) and s^2 which estimates the overall variance.

7.4.1   Application to medullary widths of cat hairs

Consider a crime in which a cat is involved. For example, in a domestic burglary, there may have been a cat at the crime scene. A suspect is identified who has cat hairs on his clothing. A full assessment of the evidence would require consideration of the suspect's explanation for the presence of these hairs and of the probabilities of transfer of cat hairs from the scene of the crime and from elsewhere. Such issues are not debated here. Measurements are made of the medullary widths, among other characteristics, of these hairs and of a sample of hairs from the domestic cat. Let $\bar{x}$ denote the mean of m hairs from the source (the domestic cat), let $\bar{y}$ denote the mean of n hairs from the receptor (the suspect's clothing). Some sample results for the value, V, as given by (7.9) of the evidence are given in Table 7.4 for various values of x, y and σ. Variation in σ is given to illustrate the effect of changes in the variation within cats of medullary width on the value of the evidence. The value of the smoothing parameter λ has been taken to be 0.09 and 0.50 to illustrate variation in V with λ. The corresponding density estimate curves are shown in Figures 7.4 and 7.5 which can be used to assess the relative typicality of the evidence.

Table 7.4   Value of the evidence for various values of $\bar{x}$ and $\bar{y}$, the smoothing parameter λ and the within cat standard deviation σ; $m = n = 10$ throughout; $s = 23$ microns

| $\bar{x}$ | $\bar{y}$ | σ | V | |
			$\lambda=0.09$	$\lambda=0.50$
15	15	10	16.50	12.01
15	25	10	1.39	0.782
15	35	10	9.81×10^{-4}	4.47×10^{-4}
110	110	10	84.48	53.61
50	50	10	6.97	6.25
50	50	16	3.86	3.93
50	50	5	16.14	12.48
50	55	10	3.75	3.54

Table 7.5   Refractive index of 2269 fragments of float glass from buildings, Lambert and Evett (1984)

Refractive index	Count	Refractive index	Count	Refractive index	Count	Refractive index	Count
1.5081	1	1.5170	65	1.5197	7	1.5230	1
1.5119	1	1.5171	93	1.5198	1	1.5233	1
1.5124	1	1.5172	142	1.5199	2	1.5234	1
1.5128	1	1.5173	145	1.5201	4	1.5237	1
1.5134	1	1.5174	167	1.5202	2	1.5240	1
1.5143	1	1.5175	173	1.5203	4	1.5241	1
1.5146	1	1.5176	128	1.5204	2	1.5242	1
1.5149	1	1.5177	127	1.5205	3	1.5243	3
1.5151	1	1.5178	111	1.5206	5	1.5244	1
1.5152	1	1.5179	81	1.5207	2	1.5246	2
1.5153	1	1.5180	70	1.5208	3	1.5247	2
1.5154	3	1.5181	55	1.5209	2	1.5249	1
1.5155	5	1.5182	40	1.5211	1	1.5250	1
1.5156	2	1.5183	28	1.5212	1	1.5254	1
1.5157	1	1.5184	18	1.5213	1	1.5259	1
1.5158	7	1.5185	15	1.5215	1	1.5265	1
1.5159	13	1.5186	11	1.5216	3	1.5269	1
1.5160	6	1.5187	19	1.5217	4	1.5272	2
1.5161	6	1.5188	33	1.5218	12	1.5274	1
1.5162	7	1.5189	47	1.5219	21	1.5280	1
1.5163	6	1.5190	51	1.5220	30	1.5287	2
1.5164	8	1.5191	64	1.5221	25	1.5288	1
1.5165	9	1.5192	72	1.5222	28	1.5303	2
1.5166	16	1.5193	56	1.5223	13	1.5312	1
1.5167	15	1.5194	30	1.5224	6	1.5322	1
1.5168	25	1.5195	11	1.5225	3	1.5333	1
1.5169	49	1.5196	3	1.5226	5	1.5343	1

7.4.2   Refractive index of glass

These data, from Lambert and Evett (1984), are shown in Table 7.5 and illustrated in Figure 7.1. There are many coincident points and an automatic choice of λ is difficult and perhaps not desirable. Figures 7.6 and 7.7 show the kernel density estimate curves for λ equal to 0.025 and 0.25. The coincidence probabilities, from equation (4.8) and values V of the evidence using (7.6) and (7.9) are given in Table 7.6.

Notice that, in general, the kernel approach leads to considerably higher values for V than does the Lindley approach. This arises from the more dispersed nature of the Lindley expression. Two examples in Table 7.6 show the failure of the

Table 7.6 Coincidence probability and value of the evidence (kernel and Lindley approaches) for various values of $\bar{x}$ and $\bar{y}$ and the smoothing parameter λ (for the kernel approach); $m = 10$, $n = 5$; within window standard deviation $\sigma = 0.00004$, between window standard deviation $\tau = 0.004$; overall mean $\mu = 1.5182$.

$\bar{x}$	$\bar{y}$	Coincidence probability	λ for kernel approach			Lindley
			0.025	0.05	0.25	
1.51500	1.51501	2.845×10^{-9}	17889	7055	2810	226
1.51600	1.51601	2.643×10^{-3}	563	489	419	191
1.51700	1.51701	2.863×10^{-2}	54.3	52.4	48.9	172
1.51800	1.51801	3.083×10^{-2}	53.3	54.4	49.2	164
1.51900	1.51901	2.246×10^{-2}	70.0	69.2	102.4	167
1.52000	1.52001	8.536×10^{-9}	5524	2297	471.2	182
1.52100	1.52101	4.268×10^{-9}	13083	4381	1397	210
1.52200	1.52201	1.321×10^{-2}	128	143	304	259
1.51500	1.51505	—	740	519	217	18.4
1.51600	1.51605	—	48.4	42.4	32.6	15.6
1.51600	1.51610	—	1.76×10^{-2}	1.74×10^{-2}	1.22×10^{-2}	6.30×10^{-3}
1.51700	1.51710	—	1.35×10^{-3}	1.42×10^{-3}	1.51×10^{-3}	5.69×10^{-3}

coincidence probability approach. These are examples in which the separation of the source ($\bar{x}$) and receptor ($\bar{y}$) fragments is such that an approach based on coincidence probabilities would declare these two sets of fragments to have come from different windows. However, both the kernel and Lindley approaches give support to the hypothesis that they come from the same window.

7.5   PROBABILITIES OF TRANSFER

7.5.1   Introduction

Consider transfer of material from the crime scene to the criminal. A suspect is found with similar material on his clothing, say. This material may have come from the crime scene. Alternatively, it may have come from somewhere else under perfectly innocent circumstances. There are two sets of circumstances to consider. First, conditional on the suspect having been present at the crime scene (C), there is a probability that material will have been transferred from the scene to the suspect. It has also to be borne in mind that someone connected with the crime may have had no fragments transferred from the scene to his person and have had fragments similar to those found at the crime scene transferred to his person from somewhere else by innocent means. Secondly, there is the probability that a person unconnected with the crime (i.e., this is conditional on a suspect not having been present at the scene, $\bar{C}$) will have material similar to the crime material on his person.

Consider the case of glass fragments as described by Evett (1986). Let $t_i (i = 0, 1, 2, \ldots)$ be the probability that, given C, i fragments of glass would have been transferred. More correctly, t_i is the probability that, given C, the presence of the suspect at the crime scene, i fragments would be found on the clothing of the suspect on searching. This allows for the mechanism of transfer and also for the mechanisms of persistence and recovery. Let b_i ($i = 0, 1, 2, \ldots$) denote the probability that a person in the relevant population will have i fragments of glass on their clothing. Evett (1986) considered two cases, one where a single fragment has been found and one where two fragments had been found. A general expression for n fragments (Evett, 1986) is given in Section 7.6.2.

7.5.2   Single fragment

The evidence E consists of three parts. The first part is the existence (m_1) of one fragment on the clothing of the suspect. The second part is that its refractive index is y. This is the transferred particle form of the evidence. Let $Pr(m_1 | \bar{C}, I)$ correspond to the proportion of people in the relevant population who have a fragment of glass on their clothing. Denote this probability by b_1. Similarly, b_0 denotes the proportion of people in the general population who do not have a fragment of glass on their clothing. The third part of the evidence is the measurements $\mathbf{x}$ on the source material. This is relevant for the determination of the numerator but not the denominator.

Consider the denominator of the likelihood ratio. This is

$$\begin{aligned} Pr(E|\bar{C}, I) &= Pr(m_1, y|\bar{C}, I) \\ &= Pr(m_1|\bar{C}, I) \times f(y|\bar{C}, I, m_1) \\ &= b_1 f(y|\bar{C}, I, m_1). \end{aligned}$$

The probability density function $f(y|\bar{C}, I, m_1)$ will be taken to be a Normal density function, with mean μ and variance τ^2 (or, more correctly, $\tau^2 + \sigma^2$) as in Section 7.2.3. Consider the numerator. If the suspect was present at the crime scene there are two possible explanations for the presence of the glass fragment on the clothing of the suspect. Either the fragment has been acquired by innocent means and no fragment has been transferred from the crime scene (an event with probability t_0) or the fragment was transferred from the crime scene and none was transferred by innocent means, an event with probability t_1. Let $\mathbf{x}$ denote the measurements on the source sample. The numerator is then

$$t_0 b_1 f(y|\bar{C}, I, m_1) + t_1 b_0 f(y|C, \mathbf{x}).$$

Notice the terms $t_0 b_1$ and $t_1 b_0$. The former is the probability that no particle is transferred from the crime scene and one particle is transferred from the background. The latter is the probability that one particle is transferred from the crime scene and no particle is transferred from the background. Note, also, that in the term involving $\bar{C}$, the fragment is assumed to have been transferred by innocent means. The probability density function for y in this situation is then the one that holds when the suspect is unconnected with the crime. Hence, the conditioning on $\bar{C}$ is permissible.

The likelihood ratio is then

$$V = t_0 + \frac{t_1 b_0}{b_1} \frac{f(y|C, \mathbf{x})}{f(y|\bar{C}, I, m_1)}, \tag{7.10}$$

(see also (5.4)). Compare this result with (7.5), derived by Grove (1980) where here t_0 replaces $(1 - p_T)$, t_1 replaces p_T, b_0 replaces $(1 - p_A)$ and b_1 replaces p_A. The ratio of the density functions $f(y|C, \mathbf{x})/f(y|\bar{C}, I, m_1)$ has been considered earlier. The extension described here accounts for possible different sources of the fragment. For the single fragment case

$$V = t_0 + \frac{t_1 b_0}{b_1} \frac{\sqrt{\sigma^2 + \tau^2}}{\sigma} \exp \left\{ \frac{(y - \mu)^2}{2(\tau^2 + \sigma^2)} - \frac{(y - \bar{x})^2}{2\sigma^2} \right\}.$$

Table 7.7 Distributional parameters for glass problems

μ	τ	α	σ
1.5186	4×10^{-3}	3×10^{-3}	4×10^{-5}

Table 7.8 Transfer probabilities for glass problems

b_0	b_1	b_2	t_0	t_1	t_2
0.37	0.24	0.11	0	0.056	0.056

Values for the distributional parameters and for the transfer probabilities from Evett (1986) are given in Tables 7.7 and 7.8. The values for b_0, b_1 and b_2 are suggested by Evett (1986) who cited Pearson *et al.* (1971). Evett contrasted these probabilities with results from Harrison *et al.* (1985) in which the numbers of people with one and two fragments on their clothing were proportionately closer. There is a need for a closer investigation of the estimation of these probabilities. The transfer probabilities are provided by Evett (1986) citing a personal communication by C. F. Candy. All these probabilities are provided primarily for illustrative purposes.

From these values $t_1 b_0/b_1 = 0.086$. Notice that $\tau/\sigma = 100$ and that $\tau^2 + \sigma^2 \simeq \tau^2$. Thus

$$\frac{\sqrt{\tau^2 + \sigma^2}}{\sigma} \simeq \frac{\tau}{\sigma}.$$

For the single fragment case

$$V \simeq 8.6 \exp \left\{ \frac{(y - \mu)^2}{2\tau^2} - \frac{(y - \bar{x})^2}{2\sigma^2} \right\}. \tag{7.11}$$

Some values for varying values of $(y - \mu)/\tau$ and $(y - \bar{x})/\sigma$ (standardised differences of y from the mean of the source fragments and from the overall mean) are given in Table 7.9. Small values of $(y - \bar{x})/\sigma$ imply similarity between source and recovered fragments. Small values of $(y - \mu)/\tau$ imply a common value of y. Notice the largest value of V is given by a small value of $(y - \bar{x})/\sigma$ and a large value of $(y - \mu)/\tau$.

7.5.3 Two fragments

The evidence E again consists of three parts. The first part is the existence of two fragments, the transferred-particle form, on the clothing of the suspect (m_2). The

Table 7.9 Some values for the likelihood ratio V for the single fragment case, from Evett (1986)

$(y - \bar{x})/\sigma$	$(y - \mu)/\tau$		
	0.0	1.0	2.0
0.0	9	14	63
1.0	5	9	38
2.0	1	2	9
3.0	0.1	0.2	0.7

second part is the measurements y_1, y_2, of their refractive indices. The third part is the bulk form **x**.

The denominator of the likelihood ratio is the probability that a person unconnected with the crime will have two fragments of glass with measurements (y_1, y_2) on his clothing. This is

$$Pr(m_2, y_1, y_2 | \bar{C}, I) = Pr(m_2 | \bar{C}, I)f(y_1, y_2 | \bar{C}, I, m_2)$$
$$= b_2 f(y_1, y_2 | \bar{C}, I, m_2).$$

For the numerator there are four possibilities as given in Table 7.10. Fragments which are categorised as 'not transferred' are assumed to have been acquired by means unconnected with the crime. These four possibilities are exclusive and the numerator can be expressed as

$$t_0 b_2 f(y_1, y_2 | \bar{C}, I, m_2) + t_1 b_1 f(y_1 | C, \mathbf{x})f(y_2 | \bar{C}, I, m_1)$$
$$+ t_1 b_1 f(y_2 | C, \mathbf{x})f(y_1 | \bar{C}, I, m_1) + t_2 b_0 f(y_1, y_2 | C, \mathbf{x}).$$

The first term corresponds to the transfer of two fragments from the background. The distribution of y_1, y_2 is independent of **x**, and may be written as shown. The other terms may be derived in a similar way.

The likelihood ratio is then

$$V = t_0 + \frac{t_1 b_1 \{f(y_1 | C, \mathbf{x})f(y_2 | \bar{C}, I, m_1) + f(y_2 | C, \mathbf{x})f(y_1 | \bar{C}, I, m_1)\} + t_2 b_0 f(y_1, y_2 | C, \mathbf{x})}{b_2 f(y_1, y_2 | \bar{C}, I, m_2)}.$$

Several assumptions are necessary in order to gain numerical insight into the variability of V. Some of these correspond to assumptions made by Evett (1986) but others differ with respect to the distributions of y and (y_1, y_2) in the single and two fragment cases, respectively.

First, it is assumed that fragments of glass from the same source have a Normal distribution with a certain mean θ and variance σ^2. This mean θ has itself

Table 7.10 Possible sources of two fragments

Fragment 1	Fragment 2	Probability
Not transferred	Not transferred	$t_0 b_2$
Not transferred	Transferred	$t_1 b_1$
Transferred	Not transferred	$t_1 b_1$
Transferred	Transferred	$t_2 b_0$

a Normal distribution with mean μ and variance τ^2 (see the earlier discussion in Section 7.2.1). When considering the two fragment case with refractive index measurements (y_1, y_2) it is assumed that the mean $\bar{y}\{= (y_1 + y_2)/2\}$ is Normally distributed, and also that the difference $\delta\{= (\text{higher} - \text{lower})\}$ of the two measurements has an exponential distribution and that these two distributions are independent.

Estimates for parameter values for these distributions based on surveys and casework are given in Evett (1986). The assumptions are as follows:

- The dimensions of the glass fragments are not relevant to the assessment of the evidential value

- The distributions of the refractive index measurements carried out on fragments from a broken window are independent in magnitude of the number present

- $(Y|\bar{C}, I, e_1) \sim N(\mu, \tau^2 + \sigma^2)$

- $f(y_1, y_2|\bar{C}, I, e_2) = f(\bar{y}, \delta|\bar{C}, I, e_2) = f(\bar{y}|\bar{C}, I)f(\delta|\bar{C}, I, e_2)$

- $(\bar{Y}|\bar{C}, I) \sim N(\mu, \tau^2 + \sigma^2/2)$

- $f(\delta|\bar{C}, I) = \alpha^{-1}\exp(-\delta/\alpha)$, an exponential distribution, see Figure 7.8

- $(Y|C, \bar{x}) \sim N(\bar{x}, \sigma^2)$

- $(\bar{Y}|C, \bar{x}) \sim N(\bar{x}, \sigma^2/2)$.

For the two fragment case, V may be written as the sum of two terms,

$$V = \phi(1) + \phi(2).$$

The first, $\phi(1)$, accounts for the case in which one fragment is transferred from the crime scene and one from the background. The second, $\phi(2)$, accounts for the case in which both fragments are transferred from the crime scene. It is assumed that the case in which no fragment is transferred from the scene has negligible probability. Assume, as usual, that τ^2 is much greater than σ^2 so that $\tau^2 + \sigma^2$ may be approximated by τ^2.

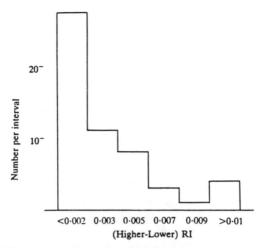

Figure 7.8   The difference in refractive index measurements (higher − lower) for each pair of fragments for individuals who had two fragments of glass on their clothing, from Harrison *et al.* (1985), (Evett, 1986) (Reproduced with permission of The Forensic Science Society.)

The general expressions for $\phi(1)$ and $\phi(2)$, using the distributional results above, can be shown to be

$$\phi(1) = \frac{t_1 b_1 \alpha \left[\exp\left\{ -\frac{(y_1 - \bar{x})^2}{2\sigma^2} - \frac{(y_2 - \mu)^2}{2\tau^2} \right\} + \exp\left\{ -\frac{(y_2 - \bar{x})^2}{2\sigma^2} - \frac{(y_1 - \mu)^2}{2\tau^2} \right\} \right]}{\sigma b_2 \sqrt{2\pi} \exp\left\{ -\frac{(\bar{y} - \mu)^2}{2\tau^2} \right\} \exp\left\{ -\frac{|y_1 - y_2|}{\alpha} \right\}}$$

and

$$\phi(2) = \frac{t_2 b_0 \alpha\tau \exp\left\{ -\frac{(y_1 - \bar{x})^2}{2\sigma^2} - \frac{(y_2 - \bar{x})^2}{2\sigma^2} \right\}}{b_2 \sigma^2 \sqrt{2\pi} \exp\left\{ -\frac{(\bar{y} - \mu)^2}{2\tau^2} \right\} \exp\left\{ -\frac{|y_1 - y_2|}{\alpha} \right\}}.$$

Two sets of results are given here in terms of $|\bar{y} - \bar{x}|/\sigma$ when $\bar{y}$ is assumed equal to μ and the parameter values and transfer probabilities are as given in Tables 7.7 and 7.8. The first set is for the case when $|y_1 - y_2| = \sigma$, the second when $|y_1 - y_2| = 4\sigma$. Let $k = |\bar{y} - \bar{x}|/\sigma$; $V = \phi(1) + \phi(2)$.

$|y_1 - y_2| = \sigma$

$$\phi(1) = 3.675 \exp\{-(4k^2 + 1)/8\}(e^{k/2} + e^{-k/2}),$$
$$\phi(2) = 566 \exp\{-(4k^2 + 1)/4\}.$$

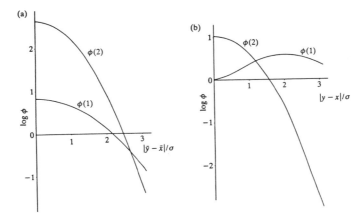

Figure 7.9 Graphs of $\log_{10}\phi(1)$ and $\log_{10}\phi(2)$ against $|\bar{y}-\bar{x}|/\sigma$ for $\bar{y}=\mu$: (a) $|y_1 - y_2| = \sigma$: (b) $|y_1 - y_2| = 4\sigma$. For the case of transfer of two fragments of glass from the scene of the crime to the criminal, the value of the evidence $V = \phi(1) + \phi(2)$ (Evett, 1986) (Reproduced with permission of The Forensic Science Society.)

$|y_1 - y_2| = 4\sigma$

$$\phi(1) = 3.825 \exp\{-(k^2 + 4)/2\}(e^{2k} + e^{-2/k}),$$

$$\phi(2) = 589 \exp(-k^2 + 4).$$

Graphs of the logarithms to base 10 of $\phi(1)$ and $\phi(2)$, plotted against $|\bar{y} - \bar{x}|/\sigma$ are shown in Figure 7.9.

Notice that the graphs of $\log_{10}\phi(2)$ are quadratic. Also, for $|y_1 - y_2| = \sigma$ the major contribution to V comes from the transfer of two fragments. For $|y_1 - y_2| = 4\sigma$, when $|\bar{y} - \bar{x}|/\sigma$ is close to 2, the major contribution to V comes from $\phi(1)$, the term corresponding to the transfer of one fragment only. This is reasonable since for $|y_1 - y_2| = 4\sigma$ and $|\bar{y} - \bar{x}| = 2\sigma$, one of y_1 or y_2 must equal $\bar{x}$.

7.5.4 A practical approach to glass evaluation

The use of probability density functions and kernel density estimation procedures is very sophisticated and requires considerable skill for their correct implementation. What is called a 'practical approach' to the interpretation of glass evidence is described by Evett and Buckleton (1990). Four scenarios are described and, for each, an expression for a likelihood ratio is derived. Sections 7.5.2 and 7.5.3 gave expressions for the value of the likelihood ratio for cases in which one or two fragments of glass have been transferred to the suspect's clothing. The four scenarios described by Evett and Buckleton (1990) concern the transfer of one or two *groups* of fragments which may or may not have come from one or two windows which have been smashed during the commission of a crime.

The circumstances are as follows. One or two windows have been smashed with criminal intent. A suspect has been apprehended very soon after the crime and one or two groups of glass fragments have been found on his clothing. The two hypotheses to be compared are:

- C: the suspect is the man who smashed the window(s) at the scene of the crime,
- $\bar{C}$: the suspect is not the man who smashed the window(s) at the scene of the crime.

Knowledge of the probabilities of various events is required. These probabilities can be estimated by reference to an appropriate clothing survey (e.g., Pearson *et al.*, 1971; Dabbs and Pearson, 1970, 1972; Pounds and Smalldon, 1978; Harrison *et al.*, 1985; McQuillan and Edgar, 1992) or by the use of personal experience. The probabilities used here are those given by Evett and Buckleton (1990). The various events with their probabilities are:

- that a person would have no glass on his clothing by chance alone, probability $b_0 = 0.636$;
- that a person would have one group of fragments on his clothing by chance alone, probability $b_1 = 0.238$;
- that a person would have two groups of fragments on his clothing by chance alone, probability $b_2 = 0.087$;
- that a group of fragments found on members of the population is large, probability $s_l = 0.029$;
- that, in the commission of the crime, no glass is transferred, probability $t_0 = 0.2$;
- that, in the commission of the crime, a large group of fragments is transferred, retained and found, probability $t_1 = 0.6$.

This set of probabilities is different from those given in Section 7.5.2. No claim is made that either set is definitive. Rather the values are given for illustrative purposes and it is a simple matter to substitute other values where this is thought appropriate. Also, the definition of large is unspecified, but again a suitable definition with an appropriate probability may be made for a particular case. If it is not felt possible to choose a particular value for a probability then a range of values may be tried. If V remains relatively stable over the range of probability values this provides reassurance that an exact value is not crucial. If V does depend crucially on the choice of a probability then careful thought is needed as to the usefulness of the method in the case under consideration.

For both windows the frequency (f_1, f_2) of occurrence of glass of the observed properties on clothing is taken to be 3%, so that $f_1 = f_2 = 0.03$ where f_1, f_2 refer to the first and second window, respectively. These frequencies may be obtained from a histogram of refractive index measurements. In a more detailed approach these values would be replaced by probability density estimates.

Four cases can be considered.

Case 1   One window is broken, one large group of fragments is found on the suspect and it is similar in properties to the broken window.

The denominator, which is derived assuming the suspect is innocent, is $b_1 s_l f_1$; i.e., the product of the probability (b_1) a person has one group of fragments on his clothing, the probability (s_l) that such a group would be large and the frequency (f_1) of glass of the observed properties on clothing.

The numerator is $b_0 t_1 + b_1 s_l f_1 t_0$. The first term accounts for the possibility that the suspect has had no glass on his clothing transferred by chance alone (b_0) and has had a large group transferred, retained and found in the commission of the crime (t_1). The probability in the circumstances that such a group of fragments has the required properties is 1. The second term accounts for the probability that the suspect has had glass of the required properties transferred by chance alone ($b_1 s_l f_1$) and no glass transferred in the commission of the crime.

The likelihood ratio is

$$V_1 = t_0 + \frac{b_0 t_1}{b_1 s_l f_1}.$$

Case 2   One window is broken and two large groups of fragments are found on the suspect. One group matches the properties of the broken window, the other does not.

The likelihood ratio is

$$V_2 = t_0 + \frac{b_1 t_1}{2 b_2 s_l f_1}.$$

The factor '2' appears in the denominator since there are two groups. The characteristics (size and frequency of occurrence) of the second group of fragments occur in both the numerator and denominator and cancel out. Hence they do not appear in the final expression.

Case 3   Two windows are broken and one large group of fragments is found on the suspect, the properties of which match one of the broken windows. The likelihood ratio is

$$V_3 = t_0^2 + \frac{b_0 t_0 t_1}{b_1 s_l f_1},$$

where it has been assumed that the transfer probabilities (t_0, t_1) are the same for both windows.

Case 4 Two windows are broken and two groups of glass are identified, one of which matches one broken window and one of which matches the other. The likelihood ratio is

$$V_4 = t_0^2 + \frac{b_0 t_1^2}{2b_2 s_I^2 f_1 f_2} + \frac{b_1 t_0 t_1}{2b_2 s_I f_1} + \frac{b_1 t_0 t_1}{2b_2 s_I f_2}.$$

Using the probability figures given above, it is easy to verify that in these four cases the first term is the dominant one. Thus, the following approximate results are obtained:

$$V_1 \simeq \frac{b_0 t_1}{b_1 s_I f_1}, \tag{7.12}$$

$$V_2 \simeq \frac{b_1 t_1}{2b_2 s_I f_1}, \tag{7.13}$$

$$V_3 \simeq \frac{b_0 t_0 t_1}{b_1 s_I f_1}, \tag{7.14}$$

$$V_4 \simeq \frac{b_0 t_1^2}{2b_2 s_I^2 f_1 f_2}. \tag{7.15}$$

The results, substituting the probability values listed into (7.12) to (7.15), are

$$V_1 \simeq 1843,$$
$$V_2 \simeq 943,$$
$$V_3 \simeq 368,$$
$$V_4 \simeq 1738\,000.$$

It would be wrong to place too much worth on the exact numerical values of these results. There are many imponderables, such as the specifications of the transfer probabilities to be considered. However, a comparison of the orders of magnitude provides a useful qualitative assessment of the relative worth of these results. For example, consider a comparison of V_3 with V_4. The latter, V_4, is bigger than the former, V_3, by a factor of about 5000. The effect on the value of the evidence, when two windows have been broken, of discovering two groups of fragments with similar properties to the broken windows (rather than just one) on the clothing of a suspect, is considerable.

7.6 APPENDIX

7.6.1 Derivation of V when the between source measurements are assumed Normally distributed.

The marginal distributions of the means of the source and receptor measurements (with m measurements on the source and n on the receptor), $\bar{X}$ and $\bar{Y}$, in the denominator are independent and are, respectively, $N(\mu, \tau^2 + \sigma^2/m)$ and $N(\mu, \tau^2 + \sigma^2/n)$.

Let $\sigma_1^2 = \tau^2 + \sigma^2/m$ and $\sigma_2^2 = \tau^2 + \sigma^2/n$ where τ^2 is the between source variance. Then $(\bar{X} - \bar{Y}) \sim N(0, \sigma_1^2 + \sigma_2^2)$ and $Z = (\sigma_2^2\bar{X} + \sigma_1^2\bar{Y})/(\sigma_1^2 + \sigma_2^2)$ is distributed as $N(\mu, \sigma_1^2\sigma_2^2/(\sigma_1^2 + \sigma_2^2))$ and $(\bar{X} - \bar{Y})$ and Z are also independent. The denominator may then be written as

$$\frac{1}{2\pi\sigma_1\sigma_2} \exp\left\{ -\frac{\cdot\, (\bar{x} - \bar{y})^2}{2(\sigma_1^2 + \sigma_2^2)} \right\} \exp\left\{ -\frac{(z - \mu)^2(\sigma_1^2 + \sigma_2^2)}{2\sigma_1^2\sigma_2^2} \right\}.$$

In the numerator, it can be shown that the joint unconditional distribution of $\bar{X}$ and $\bar{Y}$ is bivariate Normal with means μ, variances σ_1^2 and σ_2^2 and covariance τ^2.

The distribution of $\bar{X} - \bar{Y}$ is $N\{0, \sigma^2(1/m + 1/n)\}$. Let $W = (m\bar{X} + n\bar{Y})/(m + n)$. The distribution of W is $N(\mu, \tau^2 + \sigma^2/(m + n))$. Also, $(\bar{X} - \bar{Y})$ and W are independent. Let $a^2 = 1/m + 1/n$ and $\sigma_3^2 = \tau^2 + \sigma^2/(m + n)$. Then the numerator may be written as

$$\frac{1}{2\pi a\sigma\sigma_3} \exp\left\{ -\frac{(\bar{x} - \bar{y})^2}{2a^2\sigma^2} \right\} \exp\left\{ -\frac{(w - \mu)^2}{2\sigma_3^2} \right\}.$$

The value, V, of the evidence is the ratio of the numerator to the denominator and after some simplification this is

$$\frac{\sigma_1\sigma_2}{a\sigma\sigma_3} \exp\left\{ -\frac{(\bar{x} - \bar{y})^2\tau^2}{a^2\sigma^2(\sigma_1^2 + \sigma_2^2)} \right\} \exp\left\{ -\frac{(w - \mu)^2}{2\sigma_3^2} + \frac{(z - \mu)^2(\sigma_1^2 + \sigma_2^2)}{2\sigma_1^2\sigma_2^2} \right\}.$$

Large values of this provide good evidence that the suspect was at the crime scene. This expression may be simplified. Typically, τ is much larger than σ. Then $\sigma_1^2 = \sigma_2^2 = \sigma_3^2 = \tau^2$, $Z = (X + Y)/2$ and

$$V \simeq \frac{\tau}{a\sigma} \exp\left\{ -\frac{(\bar{x} - \bar{y})^2}{2a^2\sigma^2} \right\} \exp\left\{ -\frac{(w - \mu)^2}{2\tau^2} + \frac{(z - \mu)^2}{\tau^2} \right\}. \tag{7.16}$$

If the number of control measurements equals the number of recovered measurements then $m = n$, $Z = W = \frac{1}{2}(X + Y)$, $\sigma_1^2 = \sigma_2^2$ and (Lindley, 1977a).

$$V \simeq \frac{m^{1/2}\tau}{2^{1/2}\sigma} \exp\left\{ -\frac{m(\bar{x} - \bar{y})^2}{4\sigma^2} \right\} \exp\left\{ \frac{(z - \mu)^2}{2\tau^2} \right\},$$

7.6.2　Value of the evidence incorporating transfer probabilities for more than two receptor items

Consider two hypotheses:

- C: the suspect was present at the scene of the crime;
- $\bar{C}$: the suspect was not present at the scene of the crime.

The evidence E is that n fragments of glass have been found on clothing of the suspect. Measurements $\mathbf{y} = \{y_j, j = 1, \ldots, n\}$ of refractive indices have been made on these fragments. Let $\mathbf{x} = \{x_i, i = 1, \ldots, m\}$ be measurements on m source fragments from the broken window at the scene of the crime. Let T and B refer to fragments transferred from the window and to fragments transferred from the background, respectively. The evidence E can be partitioned into two parts, those fragments which came from the window (T) and those which came from the background (B). There are $N = 2^n$ such partitions. Denote the lth partition by

$$E = (E_{Tl}, E_{Bl}), \quad \mathbf{y} = (\mathbf{y}_{Tl}, \mathbf{y}_{Bl}), \quad l = 1, \ldots, N.$$

If $\bar{C}$ is true then only the partition in which all the fragments came from the background is relevant. In such circumstances the probability of the evidence is $p_\rho(E\mathbf{y}) = p_\rho(n, \mathbf{y})$ where p_ρ is the probability distribution of refractive index measurements of glass found on the clothing of members of the general population.

If C is true then consideration has to be given to the different possible partitions. Let $M_k = \binom{n}{k}$, $k = 0, 1, \ldots, n$. Thus there will be M_k partitions of the type $(k, n-k)$ where k fragments are to be compared with the source and $(n-k)$ are to be compared with the distribution of glass fragments on the clothing of members of the population. Index these partitions with the letter $s, s = 1, \ldots, M_k$. The measurements $\mathbf{y}$ of the sth partition of size k can be partitioned as $(\mathbf{y}_s^{(k)}, \mathbf{y}_s^{(n-k)})$. The distribution of the fragments $\mathbf{y}_s^{(k)}$ which are assumed to come from the crime window is a predictive distribution and can be written as

$$f(\mathbf{y}_j^{(k)} \mid k, C, \mathbf{x}),$$

where $\mathbf{x}$ denotes the refractive measurements on the fragments from the source window.　If it is also assumed that the measurements of refractive index of fragments from a broken window are independent of the number present then this expression may be written as

$$f(\mathbf{y}_j^{(k)} \mid C, \mathbf{x}).$$

From the discussion of $\bar{C}$ the distribution of the fragments $\mathbf{y}_j^{(n-k)}$ from the background population is given by

$$p_\rho(n - k, \mathbf{y}_j^{(n-k)}).$$

Let t_k denote the probability that k fragments are transferred from the crime scene to the criminal. Then, the numerator of the likelihood ratio may be written as

$$\sum_{k=0}^{n} t_k \sum_{s=1}^{M_k} f(\mathbf{y}_j^{(k)}|C,\mathbf{x})p_\rho(n-k,\mathbf{y}_j^{(n-k)}).$$

Consider r fragments, define b_r to be the probability that a person will have r fragments on his clothing and let $p_\rho^{(r)}$ denote the probability distribution of refractive indices on the clothing of people who have r fragments. Then

$$p_\rho(r\mathbf{y}^{(r)}) = p_\rho(r)p_\rho(\mathbf{y}^{(r)}|r)$$
$$= b_r p_\rho^{(r)}(\mathbf{y}^{(r)}).$$

Then the numerator of the likelihood ratio becomes

$$\sum_{k=0}^{n} t_k \sum_{s=1}^{M_k} f(\mathbf{y}_j^{(k)}|C,\mathbf{x})b_{(n-k)}p_\rho^{(n-k)}(\mathbf{y}_j^{(n-k)}).$$

and the denominator becomes

$$b_n p_\rho^{(n)}(\mathbf{y}).$$

Conventionally, the following results may be assumed:

$$f(\mathbf{y}_1^{(0)}|C\mathbf{x}) = 1,$$
$$p_\rho^{(0)}(\mathbf{y}_1^{(0)}) = 1.$$

The value of the evidence is then

$$V = t_0 + \frac{1}{b_n p_\rho^{(n)}(\mathbf{y})} \sum_{k=1}^{n} t_k b_{(n-k)} \sum_{s=1}^{M_k} f(\mathbf{y}_j^{(k)}|C,\mathbf{x})p_\rho^{(n-k)}(\mathbf{y}_j^{(n-k)}),$$

(Evett, 1986).

8

DNA Profiling

8.1 INTRODUCTION

The interpretation of DNA (deoxyribonucleic acid) profiles in forensic science has been the source of much discussion; see Roeder (1994) for a review. It is not necessary here to go into the details of the preparation of a DNA profile; the original work is documented in Jeffreys *et al.* (1985a, b) and a brief description of the technique has been given in Example 1.3. Notice though that the standard deviation of the length measurements is proportional to molecular weight. The measurement errors of a pair of bands are positively correlated; if one member of a pair of bands has a measured position slightly greater (less) than its true position then the other member will tend to have a measured position slightly greater (less) than its true position.

A single-locus probe attaches itself to one locus; the two bands observed are one from the mother, one from the father, of the person from whom the DNA was obtained. If the outcome is only a single band this may be because of homozygosity or because the two bands are sufficiently close together that the probe cannot differentiate between them (*coalesence*), or because of a null allele that escapes detection (e.g., because it is very small and runs off the gel) (Kaye, 1993b).

A multi-locus probe attaches itself to more than one locus. The outcome is many pairs of bands in a strip resembling a barcode. It is this which has been given the name *DNA fingerprint.* However, interpretation of this is much more difficult. In particular, 'good statistical methods are not available for analysing band weights of multi-locus probes' (Berry, 1991a), though see Curnow and Wheeler (1993) for an application to paternity cases. Preference is now for several single-locus probes to be used as they are more robust and reliable. The evidence combines multiplicatively, assuming linkage equilibrium, to produce the very small probabilities popularly associated with DNA analysis.

Consider the following situation in which a rape has been committed. Evidence has been found at the crime scene from which it has been possible to extract a sample of DNA (e.g., semen stains in a rape case). Also, it is possible to eliminate all innocent sources of the DNA and, hence, to claim that the DNA can only have come from the criminal. Then the following notation can be introduced:

- C: DNA in the crime sample came from the suspect;
- $\bar{C}$: DNA in the crime sample came from someone else;
- I: background information, to include reference data $D = (z_1, \ldots, z_n)$ of band positions from some relevant populations;
- E_1: Positions (x_1, x_2) of bands in DNA profile from blood sample of suspect $(x_1 < x_2)$, the source form of the evidence;
- E_2: Positions (y_1, y_2) of bands in DNA profile of crime sample $(y_1 < y_2)$, the transferred particle form of the evidence.

Let $E = (E_1, E_2)$. Notice the terminology. Samples are known as crime and suspect samples here, depending on their origin. Normally this should not cause confusion but care is needed, particularly in Section 8.6.

There are several methods of evaluating this evidence and there has been much debate about their relative merits. The discussion here is of necessity a very brief review of this debate. First, a technique known as *match-binning* for evaluating the evidence is described. This has similarities with the two-stage process for the evaluation of glass evidence described in Section 4.6. An example of the use of this technique then follows. An artificial model involving balls in an urn, with reflections back to Section 1.6, is then introduced to illustrate the effect of measurement errors on the value of the evidence. This model is a discrete model. Models which treat the band positions as continuous are then discussed, again with an example. Techniques for the incorporation of information about related individuals finishes the chapter.

8.2   MATCH-BINNING

The title, *match-binning*, of this method refers to the two stages of the analysis. The method has similarities with the coincidence probability approach of Section 4.6.

8.2.1   Matching

The crime sample and the suspect sample are examined. In general, each sample consists of a pair of bands on a vertical strip. This is so for many heterozygotes. Sometimes a single band only is visible. The problem in matching is to decide, by inspection of the positions of the two pairs of bands, whether the two samples have come from the same source. If the samples have come from the same source the positions should be similar. If the samples have not come from the same source the positions may or may not be similar.

The assessment of similarity is not easy. Samples of the same DNA will migrate different distances in different regions of the gel or on different gels. The first stage in the assessment is to whether the bands match, in some sense, or not.

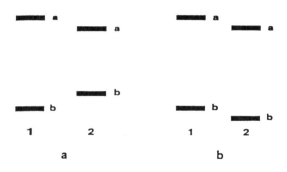

Figure 8.1 Visual matching of DNA profiles from a single-locus probe of a crime and suspect sample. The pattern in (a) would not be declared a match. The pattern in (b) might be (Budowle *et al*, 1991). (Reproduced with permission of The University of Chicago Press.)

In Figure 8.1 (Budowle *et al.*, 1991) the two patterns in (a) may be declared a match whereas those in (b) would not. A phenomenon known as band shifting may produce the difference in band positions in (b). It is unlikely to produce the discrepancy in the pattern in (a). Reproducibility studies have enabled the amount by which fragments of the same length may differ on a gel to be quantified. For example, in studies conducted by the South Carolina Law Enforcement Division (SLED) DNA Laboratory (SLED, 1992) and reported by Weir and Gaut (1993), corresponding bands never differed by more than 5.6% of their average length.

The second stage is only used if a match has been declared. In that case, the relative frequency of such a match is estimated with reference to an appropriate relevant population. This is the binning stage. Bins may be either fixed or floating.

The use of fixed bins corresponds to the use of histograms and cell frequencies, constructed without reference to the particular crime and suspect samples under consideration. If a crime or suspect band falls in a particular cell of the histogram then it is the relative frequency associated with that cell which is the relative frequency associated with that band. The width of the cells of the histogram is the same as the distance used in the matching stage.

An alternative procedure uses so-called floating bins. Floating bins are bins which are centred on the positions of the bands in the crime and suspect samples. Again, the width of the bin depends on the matching criterion, which will be such that a match is declared if the crime and suspect bands lie within a certain distance of each other. The relative frequency estimated is the proportion of bands within the relevant population which fall within this width of the extremes of the crime and suspect bands.

Suppose a match is declared if a crime band and a suspect band lie within k units of each other. Consider the positions of the first bands, x_1 and y_1, if these were compared and a match had been declared but x_1 was slightly smaller in value than y_1 then the position of the floating bin considered for these bands

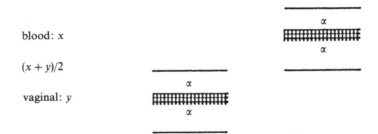

Figure 8.2 Illustration of assessment of similarity. Two bands are said to match if they differ by no more than 2α (Reproduced with permission of the American Bar Association from Weir, B. S. and Gaut, B. S. (1993) Matching and binning DNA fragments in forensic science. *Jurimetrics*, **34**, 9–19.)

would be $(x_1 - k, y_1 + k)$. Notice that a match is declared if the bands lie within k units of each other. A further k is then added in both directions to create the bin size. Thus, if two bands match exactly the width of the floating bin is $2k$; if the two bands are at the limit at which a match can be declared, i.e. k units apart, then the floating bin width will be $3k$. Notice also that heavier bands move less far through the gel. Thus for heavy bands of weight 13 to 14 *kb*, a spread of k units corresponds to a larger error in band weight than for a lighter band, of say 1.5 *kb*. For example, if $k = 1$ mm, for bands of weight 13 to 14 *kb* a 1 mm spread corresponds to an error of 4–5% in band weight, whereas for bands of 1.5 *kb*, 1 mm corresponds to an error of 1–2%.

Consider a comparison of fragment lengths from DNA obtained from vaginal swabs and from blood taken from the same woman in the rape case and consider only one band from each. Let y denote the length of the band from the DNA in the vaginal swab and let x denote the length of the band from the DNA in the blood sample from the suspect. Let α be a parameter (proportional to band size) to represent measurement error in that the *true* length of a band of *estimated* length x (y) is thought to be contained in the interval $x \pm \alpha$ ($y \pm \alpha$). Two bands are said to match if they are no more than 2α apart; see Figure 8.2.

The error is taken to be proportional to the band size. Thus

$$|x - (x + y)/2| < \alpha(x + y)/2, \tag{8.1}$$

and

$$|y - (x + y)/2| < \alpha(x + y)/2. \tag{8.2}$$

Hence, from (8.1),

$$\frac{|x - y|}{(x + y)/2} < 2\alpha,$$

and from (8.2)

$$\frac{|x - y|}{(x + y)/2} < 2\alpha.$$

The relationship

$$\frac{|x - y|}{(x + y)/2} < 2\alpha \tag{8.3}$$

defines a *similarity*. For the SLED DNA laboratory, $\alpha = 0.028$ ($= 0.056/2$).

This procedure establishes a numerical matching rule for a particular laboratory. Band lengths y from DNA from a crime scene can then be compared with band lengths x from DNA from a suspect. If the bands are declared to be similar and if both pairs of bands in the two profiles differ by no more than this criterion the two profiles are said to *match*.

Another definition of a match is based on the variation in standard deviation. Measurement error is such that the standard deviation of a single measurement is proportional to the fragment length. Thus the separation allowed for a match to be declared is greater for longer fragments. A match is then declared if the two bands fall within a certain number of standard deviations of each other or if they fall within a certain percentage of the mean weight of the two bands (one crime band, one suspect band). These two criteria can be compared since the standard deviation is proportional to the molecular weight.

The USA FBI requires that bands be separated by no more than 2.5% of the mean molecular weight (Kaye, 1993b). According to Lander (1991) this implies a standard deviation of 1.5% of molecular weight. The UK Forensic Science Service estimate one standard deviation to be 1.1% of molecular weight and use a match criterion of 2.475 standard deviations (Berry *et al.*, 1992). Despite these differences, the basic principle is that a match is declared if the bands are within a certain number (usually no more than 3) of standard deviations of each other. Mathematically, define a threshold α_i for band i ($i = 1, 2$). A match is declared between two pairs of bands if

$$|x_1 - y_1| < \alpha_1,$$
$$|x_2 - y_2| < \alpha_2.$$

Notice the dichotomy in both these methods of assessment. Two pairs of bands either match or they do not match. Consider a hypothetical 3 standard deviation criterion. Bands which were separated by 2.95 standard deviations would be given the same weight as those bands which were separated by 1.2 or 0.2 standard deviations, say. Also, a separation of 2.95 standard deviations would be recorded as a match, a separation of 3.05 standard deviations would not (Evett *et al.*, 1989a). There is a similarity here with the comparison stage of the assessment of the coincidence probability approach discussed in Section 4.6.2.

8.2.2   Floating and fixed bins

Floating bins

These are bins which are centred on the crime sample y. Any band of length x satisfying (8.3) is said to match a crime sample band of length y. An interval (or *bin*) is centred around y with a width chosen to contain such lengths. From (8.3) this bin contains all lengths x for which

$$\frac{(1-\alpha)y}{(1+\alpha)} < x < \frac{(1+\alpha)y}{(1-\alpha)}.$$

These bins are known as *floating bins*. They may be thought to float in that they are centred about the band positions of the crime sample, rather than being fixed in advance of, and in ignorance of, the positions of the crime samples. The widths of these bins are

$$\left\{ \frac{(1+\alpha)}{(1-\alpha)} - \frac{(1-\alpha)}{(1+\alpha)} \right\} y = \frac{4\alpha y}{(1-\alpha^2)} \simeq 4\alpha y$$

if α is assumed sufficiently small that α^2 may be considered negligible compared with 1 {e.g., with $\alpha = 0.028$ (SLED, 1992), $\alpha^2 < 0.0008$}. Thus the widths of the bins may be considered to be ($4\alpha \times$ the length of the crime sample). The frequencies of occurrences of band lengths from some relevant population in these windows can then be determined with reference to the database D.

Fixed bins

A method using fixed bins, in which a set of bins were defined with fixed boundaries before any measurements were made, was introduced by the FBI (Budowle *et al.*, 1991). Following the declaration of a match a window of width 2α, centred on y, the length of the crime sample band, is constructed. If this window lies entirely within a fixed bin, the band is assigned to that bin, and the frequency used in the construction of an overall profile frequency is the frequency, within the database, associated with that bin. If the window includes a bin boundary, the band is assigned to the bin with the higher frequency. Such a procedure results in the highest value for the profile frequency. It is thus the most favourable for the suspect.

   A simplistic approach to the estimation of the likelihood ratio takes the numerator, the probability of finding the evidence on the guilty person, to be 1 and the denominator of the likelihood ratio, the probability of finding the evidence on an innocent person, to be the profile frequency. Thus, the higher the profile frequency, the lower the value of V.

Note that the analysis is based on a scene-anchored perspective. If a match is declared, its significance is assessed under the assumption that the suspect did not leave the crime scene evidence. The assessment therefore is independent of the suspect. As stated in Section 5.3.1 the ethnicity of the suspect is irrelevant (see also Weir and Evett, 1992; Evett and Weir, 1992). The band lengths in the suspect profile are irrelevant. If both x and y were used in the construction of the bins this could result in a bin of width 6α for the floating bin method and of width 3α for fixed bins (Weir and Gaut, 1993).

Once a match has been declared, the binning stage is considered. Whether fixed or floating bins are considered, the proportion of bands in D which fall in the appropriate bin is noted. For $i = 1, 2$, let this proportion be p_i. Then, the proportion of people matching the crime sample is taken to be $2p_1p_2$. The factor 2 arises because one band comes from the father, one from the mother and it is not known which is which; thus there are two possible ways in which the particular band pattern could have arisen. The product p_1p_2 is justified on the basis of Hardy–Weinberg equilibrium.

Consider, in general, a profile with m loci, consisting of m pairs of different lengths e_i ($i = 1, \ldots, 2m$), obtained by using m different probes, each probe being associated with a different locus on the chromosome, and corresponding floating bin frequencies p_i ($i = 1, \ldots, 2m$). The overall frequency associated with the profile may be determined, as above, using assumptions of Hardy–Weinberg equilibrium (to multiply frequencies within loci) and linkage equilibrium (to multiply frequencies between loci). The product of the frequencies within each locus has also to be further multiplied by two. The overall profile frequency is then estimated by

$$P = 2^m \prod_{i=1}^{2m} p_i \tag{8.4}$$

(Weir and Gaut, 1993).

It is this multiplicative process which gives rise to the very small probabilities associated with DNA profiling. If only one probe is involved, the probabilities involved may be of the order of $1/50$. However, if four such probes are involved, the proportions are of the order of $(1/50)^4$ or $1/(6.25 \times 10^6)$.

Various comments need to be made. First, if a single band is seen at a locus, standard practice is to assume it is an apparent homozygote and use twice the frequency of that bin for the contribution of the locus to the profile frequency. If it were thought to be a true homozygote the square of the frequency would be used. Also, direct counting of the number of matches in a database with the profile derived from the evidence is not feasible since no database profile has been found to match an evidence profile at all bands; hence the product rule for the estimation of a profile frequency is used. Finally, the independence assumptions have been the cause of much of the controversy surrounding the use of DNA profiling.

An introduction to the debate may be found in Lander (1989), Cohen (1990), Weir (1992) and Risch and Devlin (1992). It has been commented that when statistical tests were used to try to detect population substructure the results were 'virtually meaningless because the tests have such low statistical power to detect substructure even if it is present' (Lander, 1991). However, it has also been remarked that 'whatever levels of dependence do exist are unlikely to have a meaningful impact on forensic calculations' (Weir, 1992).

8.2.3   Big bins and guessing

Two alternative assessment procedures to circumvent the criticisms of the assumptions concerning Hardy–Weinberg and linkage equilibrium have been noted (Kaye, 1993b).

The first is to consider bins with a width wider than the match windows, so-called *big bins*. This procedure provides a conservative estimate in that the relative frequency is larger than it should be to reflect the true relative frequency in the population. As such it favours the defendant since it implies the DNA profile is less rare than if the true frequency had been used.

The second assessment procedure has been called *guessing* (Kaye, 1993b). In one case reported by Kaye (1993b), the method of assessment using independence with big bins provided a figure of 1 in 67 million. The court asked the geneticist who provided this figure to provide, instead, 'the most conservative possible estimate conceivable'. This was 1 in 100 000 and it was this figure which was held to be admissible.

8.2.4   Ceiling principle

In 1992, a committee of the National Research Council (NRC, 1992) of the US National Academy of Sciences, the Committee on DNA Technology in Forensic Science, published a report of the same name. This report was a response to increasing concern over DNA profiling. One of the concerns related to the use of the multiplication rules for combining the frequencies of individual alleles, for Hardy–Weinberg and for linkage equilibrium as discussed above. These rules are based on the assumption that the relevant population does not contain sub-populations with distinct allele frequencies. Serious concern, however, has been expressed among population geneticists about the possibility of significant sub-structure and its effect (Lewontin and Hartl, 1991). However, other population geneticists have concluded that even if there were a population substructure, the evidence to date is that it has only a minimal effect on estimates of genotype frequencies (Chakraborty and Kidd, 1991). It was to resolve this conflict of opinions that the ceiling principle was proposed. The ceiling principle, as described by the NRC report (1992, pp. 83, 84, 91, 92) is as follows.

Random samples of 100 persons should be drawn from each of 15–20 populations, each representing a group relatively homogeneous genetically. The largest allele frequency in any of these populations, or 5%, whichever is larger, is to be taken as the 'ceiling' frequency for that allele. Results from different single-locus probes can then be multiplied together to give the appropriate frequency for assessing the value of the DNA profile. The Committee claim that this approach is independent of subpopulation structure, is conservative in nature in that it favours the suspect and is a sensible procedure to adopt. However, it errs in concentrating on the need for conservative estimation of relative frequencies while ignoring the point that the matching stage has also to be carried out efficiently (Evett *et al.*, 1993).

The Committee proposes an even more conservative approach until ceiling frequencies can be estimated from population studies. For each allele, a modified ceiling frequency is to be determined by calculating the 95% upper confidence limit for the allele frequency in each of the existing population samples and using the largest of these values, or 10%, whichever is larger. Once the ceiling for each allele is determined, the multiplication rule is applied. These conservative values can be replaced by those previously described as appropriate population surveys are completed.

These suggestions have been widely criticised (see, for example, Devlin *et al.*, 1993), who cite Evett *et al.* (1993) and Devlin and Risch (1992) to state that 'even if ethnic databases are composed of a mixture of subpopulations, the effects on estimates of genotype probability will be small'. Devlin *et al.* (1993) also criticise the sampling procedures, which suggested 100 people for a subpopulation, as providing samples which are too small for the estimation of allele frequencies. Also, the populations suggested by the Committee include English, Germans, Italians, Russians, Navahos, and Puerto Ricans. However, for a crime committed in Boston, USA, Devlin *et al.* (1993) suggest that information about allele frequencies in Navahos may not be relevant. Also, the arbitrary values of 5% and 10% are difficult to justify, especially since some alleles have been found to be rare in a wide range of populations. Again, such techniques may not be necessary in somewhere like Scotland where the population is overwhelmingly from one ethnic group (Caucasian). The Committee's suggestion of the multiplication rule for confidence limits can also be criticised since they suggest using the product of confidence limits but this is not the confidence limit of the product which is what is required.

Recent correspondence (Lewontin, 1993; Weir and Evett, 1993) exemplifies the debate which continues regarding the determination of the population which is appropriate for the estimation of relative frequencies. Both Lewontin (1993) and Weir and Evett (1993) agree that the ethnic background of the suspect has no relevance to the calculations used in the forensic uses of DNA. Weir and Evett (1993) differ, however, with Lewontin (1993) when Lewontin argues that the evidence is unsatisfactory because the question of which population is relevant, and hence the question of what relative frequency to assign to a particular bin in the match-binning method, cannot be answered. Weir and Evett (1993) argue that the 'discriminating power of DNA profiles is so high and the chance of a fortuitous match so low that the precision with which a frequency is estimated is rarely, if ever, going to have an unnecessarily prejudicial effect on the decisions of a court'.

Notice that in Chapter 6, the differences in value which may arise from consideration of general and particular populations were discussed in Sections 6.2.2 and 6.2.3. Here, however, it is argued that, while there may be population differences, the forensic significance is not too great. This may seem paradoxical. It is not, since in both situations the relevant population has still to be chosen as carefully as possible. It is desirable that if a particular ethnic subgroup is identified as the relevant population the information about that subgroup should be used where possible. When such a subgroup has not been identified then information from a general population should be used.

8.3   EXAMPLE OF MATCH-BINNING (SOUTH CAROLINA v. REGISTER)

An example of the use of the match-binning procedure is given by Weir and Gaut (1993) with reference to the case of *State v. Register* (Case number 38918001, State of South Carolina, County of Horry). Johnnie K. Register II was convicted in January 1993 by a South Carolina jury of rape and homocide. Semen samples taken from the victim and her clothing gave profiles which were found to match that of Register. There were three samples from the crime scene and one from the suspect, Register. The results for probe D1S7 are given in Table 8.1, extracted from Weir and Gaut (1993) where results for other probes (D2S44, D4S139, D10S28, all two bands, and D17S79, one band) are also given. A check for matching can be made using $\alpha = 0.028$ and (8.3). Consider, for example, band 2 and compare the suspect sample (x) with crime sample 2 (y). Then, $x = 9019$, $y = 9089$, $|x - y| = 70$, $(x + y)/2 = 9054$, $2\alpha = 0.056$ and

$$\frac{|x - y|}{(x + y)/2} = 0.0099 < 0.056.$$

Table 8.1   Fragment lengths in base pairs for D1S7 for the Register case

Band	Crime sample			Suspect sample
	1	2	3	
1	9024	9089	9079	9019
2	2166	2170	2162	2164

Table 8.2 Bin frequencies for the Register case

Probe	Length (bp)	Frequencies			
		Caucasian		Black	
		Fixed bin	Floating bin	Fixed bin	Floating bin
D1S7	9064	0.096	0.093	0.033	0.024
	2166	0.025	0.019	0.035	0.041
D2S44	1940	0.086	0.124	0.063	0.094
	1674	0.131	0.175	0.139	0.102
D4S139	7453	0.175	0.095	0.118	0.087
	4352	0.050	0.059	0.110	0.100
D10S28	3087	0.056	0.066	0.043	0.055
	2943	0.056	0.082	0.065	0.082
D17S79	1761	0.168	0.171	0.055	0.057

For the calculation of bin frequencies, an average of the three crime samples was used for each band. Fixed bin and floating bin frequencies for the nine distinct bands in the profile are given in Table 8.2, extracted from Weir and Gaut (1993) for Caucasian and for Black populations in South Carolina.

Frequencies for the crime scene profile can be estimated from Table 8.2. For a South Carolina Caucasian population the frequency for fixed bins is, from (8.4) with $m = 5$ and an adjustment for an apparent homozygote, $0.096 \times 0.025 \times 0.086 \times 0.131 \times 0.175 \times 0.050 \times 0.056 \times 0.056 \times 0.168 \times 2^5 \simeq 1/250$ million. For floating bins, the frequency is $1/157$ million. For South Carolina Blacks, the frequencies are approximately $1/1500$ million for both fixed and floating bins.

8.4 AN URN MODEL AND LIKELIHOOD RATIOS

8.4.1 Construction of the model

It is possible to develop an approach to estimate the value of DNA profiles in evidence based on likelihood ratios. The hypotheses to be considered are

- C: the crime and suspect samples came from the same source;
- $\bar{C}$: the crime and suspect samples did not come from the same source.

The evidence E is the band positions of the DNA profiles of the two samples. Then, the value V of the evidence is given by

$$V = \frac{Pr(E|C)}{Pr(E|\bar{C})}.$$

There is debate concerning the best approach to the estimation of V. See, for example, Gjertson *et al.*, (1988), Morris *et al.* (1989) (both for paternity), Berry (1991a), Berry *et al.* (1992), Devlin *et al.* (1992), Evett *et al.* (1992b), Jarjoura *et al.* (1994). There is also debate regarding the use of likelihood ratios in the context of DNA profiling. For example, the NRC (1992, p.85) report commented that '... outside the field of paternity testing, no forensic laboratory in this country (USA) has, to our knowledge, used Bayesian methods to interpret the implications of DNA matches in criminal cases'.

The Report also comments earlier that 'no testing laboratories in the United States now use that (likelihood ratio) approach. The Committee recognises its intellectual appeal but recommends against it. Accuracy with it requires detailed information about the joint distribution of fragment positions and it is not clear that information about a match could be understood easily by lay persons' (NRC, 1992, p.62). However, Kaye (1993b) has remarked that 'none (of the likelihood ratio methods suggested) can be dismissed as unreasonable or based on principles not generally accepted among the statistical community. Therefore, as with match frequencies, unless likelihood ratios are so unintelligible as to provide no assistance to the jury or so misleading as to be unduly prejudicial, they should be admissible.'

There are two main suggestions for the derivation of a likelihood ratio. One approach (Devlin *et al.*, 1992) considers the allele positions as discrete and use a deconvolution approach. A simple case of their model is introduced below to illustrate the effect of measurement error on the value of V but a full analysis of their model is beyond the scope of this book. The other approach (Berry, 1991a; Berry *et al.*, 1992; Evett *et al.*, 1992b; Jarjoura *et al.*, 1994) considers a convolution approach in which the allele positions are taken to be continuous and there is a convolution of the distribution of their positions with the Normal distribution associated with the measurement errors.

After the discussion of the example from Devlin *et al.* (1992) to illustrate the effect of measurement errors, the approach of Berry (1991a) which accounts for sampling errors as well as measurement errors is described. The approach uses a univariate model which assumes the allele positions to be continuous and considers each band position in isolation. It is fairly straightforward and requires only a few lines of computer code to implement. It provides a good illustration of the problems associated with the estimation of the value of a DNA profile.

The univariate model of Berry (1991a) assumes the measurement errors are independent. However, there is considerable evidence in support of a positive correlation r in the measurement errors of the two bands of a single-locus probe (Evett *et al.*, 1989a). For example, Berry *et al.* (1992) quote $r = 0.904$ for YNH24, Berry (1991a) quotes $r = 0.33$ for DS244 and $r = 0.70$ for D17S79. Berry *et al.* (1992) extend the ideas of Berry (1991a) by considering a bivariate density estimation problem but a full development of these models is beyond the scope of this book.

In Section 1.6 an analogy involving balls in an urn was used to develop a standard for probability. Another analogy with balls in an urn can be used to illustrate how measurement errors complicate the issue of assessing the value of the evidence of a DNA profile (Devlin *et al.*, 1992).

Consider a number of balls (the population) in an urn. They are of three different colours, red (R), green (Gn) and grey (Gy), corresponding to three different alleles. The different colours of balls have different weights, corresponding to the measured fragment lengths. The balls when drawn from the urn are to be observed by a colour-blind individual H who cannot distinguish the balls by colour but only by weight. This is analogous to the situation with DNA profiling where alleles can only be distinguished by measurements of their fragment lengths. If the weights can be determined accurately then the inference of a colour from a weight will be made exactly. However, if there are measurement errors, which may arise, for example, if H has to judge the weight of the ball by holding it in his hand, then an incorrect inference may occur.

A ball is selected at random from the urn by an independent selector, unseen by H. All balls are equally likely to be selected. The weight, x, of the selected ball is measured with error by H. The ball is then returned to the selector. Then, with a certain, unknown probability the selector returns the ball to H; otherwise he gives H a different ball which has been selected, again at random, from the urn. H then makes a second measurement y of weight, again with error, and independently of the first measurement, but H does not know whether this is of the original ball or of a different ball.

Two hypotheses are to be compared, with the analogous hypothesis from DNA profiling in parentheses:

- C: the same ball was measured twice (the DNA profile of the same individual was obtained twice);
- $\bar{C}$: two different balls were measured (the DNA profiles of two different individuals were obtained).

The population of balls in the urn is such that 80% are red (R), 15% are green (Gn) and 5% are grey (Gy). At any one draw by the selector, the probabilities that a red, green or grey ball is drawn are

$$Pr(R) = 0.80, Pr(Gn) = 0.15, Pr(Gy) = 0.05;$$

$$Pr(R) + Pr(Gn) + Pr(Gy) = 1.$$

The weights x of the balls are taken to be discrete and to take values from the first five positive integers 1, 2, 3, 4 and 5 units of weight. The probability function for the weights is conditional on the colour, which can take three possible values. Denote a colour by i and assign to i one of three possible values 1, 2 or 3 corresponding to whether the ball in question is red, green or grey, respectively. Denote the conditional probability function for a particular weight x, given

Table 8.3 Conditional probability distribution of weights of balls in an urn, given colour, when weights are measured with uncertainty, and the probability distribution of colour, after Devlin *et al.* (1992)

				Weight			
i	Colour	1	2	3	4	5	p_i
1	Red	0.25	0.50	0.25	0	0	0.80
2	Green	0	0.25	0.50	0.25	0	0.15
3	Grey	0	0	0.25	0.50	0.25	0.05

a particular colour, by $g_i(x)$ such that

$$g_i(x) = Pr(Weight = x | Colour = i); \ x = 1, \ldots, 5; \ i = R, \ Gn, \ Gy.$$

Denote the probability function for a particular colour i by p_i and then

$$p_1 = Pr(R) = 0.80, \ p_2 = Pr(Gn) = 0.15, \ p_3 = Pr(Gy) = 0.05.$$

Suppose the conditional distribution $g_i(x)$ of the weights, given the colours, is given by Table 8.3 (Devlin *et al.*, 1992). The conditional probability distribution may be thought of as a representation of the situation in which the true weights of the balls are red $= 2$, green $= 3$ and grey $= 4$ but that there is a possible measurement error of ± 1 associated with each ball. The (unconditional) probability distribution $f(x) = Pr(Weight = x)$ (so that, for example, $f(2)$ is used to denote the probability that the weight of a ball drawn from the urn is 2 units) of the weight of a ball drawn from the urn can be calculated from the figures given in Table 8.3, using the law of total probability (1.9). Thus

$$f(x) = Pr(\text{Weight} = x | \text{Colour} = \text{Red})Pr(\text{Red})$$
$$+ Pr(\text{Weight} = x | \text{Colour} = \text{Green})Pr(\text{Green})$$
$$+ Pr(\text{Weight} = x | \text{Colour} = \text{Grey})Pr(\text{Grey})$$
$$= g_1(x)p_1 + g_2(x)p_2 + g_3(x)p_3$$
$$= \sum_{i=1}^{3} g_i(x)p_i.$$

The probability distribution $f(x)$ of the weights of the balls drawn from the urn is given in Table 8.4. The weights x, y measured provide the evidence to be evaluated (corresponding to the measured DNA profiles). The hypotheses $C, \bar{C}$ are to be compared (corresponding to x, y coming from the same or from different

Table 8.4 Probability distribution of weights of balls drawn from the urn when the weights are measured with uncertainty

Weight(x)	1	2	3	4	5	Total
Probability $\{f(x)\}$	0.2000	0.4375	0.2875	0.0625	0.0125	1

individuals). The value V of the evidence for comparing the two hypotheses is then

$$V = \frac{Pr(x, y \mid C)}{Pr(x, y \mid \bar{C})}.$$

Consider the numerator first. Hypothesis C is assumed true, the same ball has had its weight measured twice. Thus

$$Pr(x, y \mid C) = \sum_{i=1}^{3} Pr(x, y \mid i, C) Pr(i \mid C)$$

$$= \sum_{i=1}^{3} Pr(x, y \mid i, C) Pr(i)$$

since the colour i of the ball does not depend on it being measured twice (similarly the DNA profile of a person does not depend on whether he is a criminal or not—criminality is assumed to be not genetic). The two measurements x, y are independent measurements on the same ball so

$$\sum_{i=1}^{3} Pr(x, y \mid i, C) Pr(i) = \sum_{i=1}^{3} Pr(x \mid i, C) Pr(y \mid i, C) Pr(i)$$

$$= \sum_{i=1}^{3} g_i(x) g_i(y) p_i$$

since $g_i(x)$, $g_i(y)$ do not depend on C either.

Consider the denominator. Hypothesis $\bar{C}$ is assumed true. Two different balls have each had their weight measured once (two unrelated individuals have had their DNA profile measured). Thus

$$Pr(x, y \mid \bar{C}) = Pr(x \mid \bar{C}) Pr(y \mid \bar{C})$$

$$= f(x) f(y),$$

and

$$V = \frac{\sum_{i=1}^{3} p_i g_i(x) g_i(y)}{f(x) f(y)}, \tag{8.5}$$

a discrete version of (7.3).

8.4.2   Evaluation of the likelihood ratio

No measurement error

First, assume that there is no measurement error. This provides results against which the effect of measurement error can be compared. Tables 8.3 and 8.4 assume there is measurement error so they are not applicable at present.

Since there is no measurement error, it is possible given the colour of the ball to predict its weight with certainty. Balls of colour red, green and grey have weights 2, 3 and 4, respectively, so

$$g_1(2) = 1, g_2(3) = 1, g_3(4) = 1.$$

Also,

$$g_1(x) = 0, x \neq 2;$$
$$g_2(x) = 0, x \neq 3;$$
$$g_3(x) = 0, x \neq 4.$$

In particular,

$$g_i(x)g_i(y) = 0 \quad \text{for} \quad i = 1, 2, 3; x \neq y.$$

If $x \neq y$ then $V = 0$ since if $g_i(x) \neq 0$ then $g_i(y) = 0$. Thus, if there is no measurement error and the two balls have different weights then they are of different colours. (If there is no measurement error and the two DNA profiles are different then the sources of the two profiles are two different individuals.)

Suppose $x = y = j$ where j takes one of the values 2, 3 or 4. Then the numerator of V is given by

$$\sum_{i=1}^{3} p_i g_i(x)g_i(y) = p_1 \quad \text{if } j = 2;$$
$$= p_2 \quad \text{if } j = 3;$$
$$= p_3 \quad \text{if } j = 4.$$

When the weights are measured with certainty then the probability distribution of the weights of the balls is just the same as the probability distribution of the colours of the balls. For example, the probability that a ball is of weight 2 is the same as the probability that p_1 a ball is red. Thus, suppose x and y both equal j, then

$$f(x) = f(y) = f(j) = p_{(j-1)}, j = 2, 3, 4,$$

and the numerator takes the value $f(j)$. The denominator, $f(x)f(y)$ equals $\{f(j)\}^2$.

The value of the evidence is then

$$V = \frac{1}{f(j)}, \; j = 2, 3, 4,$$

$$= \frac{1}{p_1} \text{ if the colour is red,}$$

$$= \frac{1}{p_2} \text{ if the colour is green,}$$

$$= \frac{1}{p_3} \text{ if the colour is grey.}$$

This result corresponds to the value of the evidence for blood groups described in Section 6.2.2. For $p_1 = 0.80$, $p_2 = 0.15$, $p_3 = 0.05$, $V = 1.25, 6.67, 20.00$, respectively.

Measurement error present

The results in Tables 8.3 and 8.4 are now applicable. Tables of probabilities for the numerator $\{\sum_{i=1}^{3} p_i g_i(x) g_i(y)\}$ and denominator $\{f(x) f(y)\}$ can be calculated for the 25 possible combinations of weights (x, y) for the two measurements and the results are given in Tables 8.5 and 8.6. Combining Tables 8.5 and 8.6 by taking the ratios of corresponding cell entries gives the values in Table 8.7 for the value of the evidence (x, y). These results may be interpreted as follows. Consider, for example, $x = y = 5$. Then $V = 15.50$. The evidence is 15.50 times more likely if the weights are weights of the same ball than if they were weights of different balls. This is a relatively high figure since only grey balls give weights of 5 and only 5% of balls in the population are grey.

Consider $x = 1, y = 3$. Then $V = 0.87$. The evidence lends slight weight to the hypothesis that the two measurements are weights of different balls; see

Table 8.5 Values of the numerator in (8.5) when there is measurement error in the weights x, y

Weight x	Weight y				
	1	2	3	4	5
1	0.0500	0.1000	0.0500	0	0
2	0.1000	0.2094	0.1188	0.0094	0
3	0.0500	0.1188	0.0906	0.0250	0.0031
4	0	0.0094	0.0250	0.0219	0.0062
5	0	0	0.0031	0.0062	0.0031

Table 8.6 Values of the denominator in (8.5) when there is measurement error in the weights x, y

Weight x	Weight y				
	1	2	3	4	5
1	0.0400	0.0875	0.0575	0.0125	0.0025
2	0.0875	0.1914	0.1258	0.0273	0.0055
3	0.0575	0.1258	0.0827	0.0180	0.0036
4	0.0125	0.0273	0.0180	0.0039	0.0008
5	0.0025	0.0055	0.0036	0.0008	0.0002

Table 8.7 Value of the evidence in support of the hypothesis C that the two measurements were of the same ball (two DNA profiles were from the same source) when there is measurement error in the weights x, y

Weight x	Weight y				
	1	2	3	4	5
1	1.25	1.14	0.87	0	0
2	1.14	1.09	0.94	0.34	0
3	0.87	0.94	1.10	1.39	0.86
4	0	0.34	1.39	5.62	7.75
5	0	0	0.86	7.75	15.50

Table 8.8 The value of the evidence $x = y$ when there is not and when there is measurement error

	Weight x, y		
	1	2	3
No measurement error	1.25	6.67	20.00
Measurement error	1.09	1.10	5.62

Table 2.9. The values of V when there is not and when there is measurement error are given in Table 8.8. Such a comparison is only possible when the weights are equal. Notice that the effect of the measurement error is to reduce the value of the evidence of a match between the two measurements. This is intuitively reasonable.

The prior odds in favour of C may be incorporated in the result to determine the posterior odds in favour of C. Let E denote the evidence (x, y) of the weights. The prior odds in favour of C is $Pr(C)/Pr(\bar{C})$. The posterior odds in favour of C, given E, is then

$$\frac{Pr(C|E)}{Pr(\bar{C}|E)} = V \times \frac{Pr(C)}{Pr(\bar{C})}$$

from (2.9). It is possible to determine $Pr(C|E)$ as

$$Pr(C|E) = \frac{V \times Pr(C)}{V \times Pr(C) + Pr(\bar{C})}.$$

From this result it can be seen that for the colour-blind individual H to determine the probability that the same ball was indeed weighed twice it is necessary to know the prior probability $Pr(C)$ of the same ball being selected twice originally, before the evidence of the weights was available. This probability may not necessarily be available to H; a selector was used to choose the balls to be handed to H.

This situation has a direct analogy with the assessment of two DNA profiles. The scientist, who is taking the role of H, evaluates V. However, to determine the effect of the evidence E on the probability of C, the hypothesis that the two profiles are from samples from the same person, i.e., $Pr(C|E)$, knowledge is required of the prior probability $\{Pr(C)\}$ that they came from the same person. This knowledge is not normally available to the scientist. The jury have to make some assessment of it on the basis of all the other information I available to them at the time of the assessment of E. For this reason it is very dangerous for a scientist to make statements concerning the probability that two DNA samples came from the same person.

8.5 LIKELIHOOD RATIO FOR CONTINUOUS DATA—SINGLE BAND CASE

Let the DNA evidence be the measurements of a single band weight for a crime scene sample (y) and a suspect sample (x). There are also available band weights $D = (z_1, \ldots, z_n)$ for n other individuals from some relevant population. These measurements are all subject to error. The hypotheses to be compared are:

- C: the suspect and crime sample came from the same person.
- $\bar{C}$: the suspect and crime sample came from different people.

The likelihood ratio V is then the ratio of the likelihood of $(x, y, z_1, \ldots, z_n | C)$ to the likelihood of $(x, y, z_1, \ldots, z_n | \bar{C})$. As the measurements are all continuous, these likelihoods may be represented by probability density functions which will here be denoted f. Thus

$$V = \frac{f(x, y, z_1, \ldots, z_n | C)}{f(x, y, z_1, \ldots, z_n | \overline{C})}.$$

Following Berry (1991a) it is assumed that the measurement errors are distributed lognormally with constant standard deviation c. (The errors are said to have a *lognormal distribution* if the natural logarithms of the errors have a Normal distribution.)

Following the notation of Section 3.4.1,

$$\log y \sim N(v, c^2),$$
$$\log x \sim N(\mu, c^2),$$
$$\log z_i \sim N(\mu_i, c^2), \qquad i = 1, \ldots, n.$$

For example, if $\log y$ is Normally distributed, the probability density function $f(y)$ of y is

$$f(y) = \frac{1}{cy\sqrt{2\pi}} \exp\left\{ -\frac{1}{2}\left(\frac{\log y - v}{c}\right)^2 \right\}.$$

Baird *et al.* (1986) have shown that the standard deviation of the measurement errors is proportional to the band weight. By considering logarithms of band weights this variation in the standard deviation is removed (see Section 8.9.1) and it is mathematically convenient to consider logarithms. Berry (1991a) has noted from data provided by Baird that they are supportive of the assumption that measurement errors are Normally distributed and, also, of the assumption that they are lognormally distributed with constant standard deviation. The standard deviation is small so these two assumptions are similar.

Suppose η is the true weight of the crime sample band and that the mean of the distribution of y is η, then it can be shown that v, the mean of the distribution of $\log y$ is equal to $\log \eta - c^2/2$ (note that the mean of the distribution of the logarithm of a weight is not the logarithm of the mean of the distribution of the weight). Let m and s be the mean and standard deviation of the two observations $\log x$ and $\log y$. Thus

$$m = \tfrac{1}{2}(\log x + \log y) = \log(xy)/2,$$
$$s^2 = [\{\log x - \log(xy)/2\}^2 + \{\log y - \log(xy)/2\}^2]/2$$
$$= [\{\tfrac{1}{2}(\log x - \log y)\}^2 + \{\tfrac{1}{2}(\log y - \log x)\}^2]/2$$
$$= (\log x - \log y)^2/4,$$
$$s = \frac{1}{2}\left|\log\left(\frac{x}{y}\right)\right|.$$

Similarly, s_i, the standard deviation of $\log z_i$ and $\log y$ is

$$s_i = |\log(z_i/y)|/2 \ (i = 1, \ldots, n).$$

8.5.1 Population case

If C is true, $\mu = v$. Assume, rather unrealistically, that the probability that a band weight lies in a particular interval is proportional to the length of that interval; in other words the distribution of $\{\mu_i, i = 1, \ldots, n\}$ is said to be *uniform*. The numerator of the likelihood ratio is then $f(x, y, z_1, \ldots, z_n | \mu = v)$ which is the integral of $f(x, y, z_1, \ldots, z_n | \mu = v, \mu, \mu_1, \ldots, \mu_n)$ integrated over all possible values of $\mu, \mu_1, \ldots, \mu_n$, assuming a uniform distribution for $\{\mu_i, i = 1, \ldots, n\}$. This is a continuous version of the law of total probability for discrete events (Sections 1.6.6 and 7.1). The calculation assumes that $\{\mu, \mu_i, i = 1, \ldots, n\}$ is the population of band weights and that there is no sampling error. There is, however, still measurement error. Also, the measurements x and y are independent when $\mu = v$ and the measurements $z_1, \ldots, z_n$ are mutually independent of each other and of x and y. Thus, it can be shown (see Section 8.9.2) that the likelihood ratio V obtained by assuming that the database contains the whole population of size n and that there is therefore no sampling error is

$$V = \frac{\exp\{-(s/c)^2\}}{n^{-1}\sum_{i=1}^{n}\exp\{-(s_i/c)^2\}}, \tag{8.6}$$

(Berry, 1991a). Notice that the maximum of this ratio occurs when $x = y$ so that $s = 0$ and the numerator takes the value 1. Variation in the values of V as proportions of the maximum value of V, V_{max} say, as the standard deviation of $\log x$ and $\log y$ varies is illustrated in Table 8.9.

8.5.2 Sample case

Suppose, now, that the data $D = \{z_i; i = 1, \ldots, n.\}$ are a representative sample from an underlying population. Sampling errors have then to be considered. This

Table 8.9 Values of V as proportions of V_{max}

$\log x - \log y$	$s = (\log x - \log y)/2$	V
0	0	V_{max}
$2c$	c	$e^{-1} V_{max} = 0.368\, V_{max}$
$4c$	$2c$	$e^{-4} V_{max} = 0.018\, V_{max}$
$6c$	$3c$	$e^{-9} V_{max} = 0.00012\, V_{max}$

is done by representing the variation in μ as a conditional probability density function where the conditioning is on the values contained in D. The likelihood ratio, V, for a single band is then

$$V = \frac{f(x, y | \mu = v, D)}{f(x, y | D)}.$$

The numerator

$$f(x, y | \mu = v, D) = \int f(x, y | \mu, v, \mu = v) f(\mu | D) d\mu$$

$$= \int \frac{1}{2\pi c^2 xy} \exp\left[-\frac{1}{2c^2} \{ (\log x - \mu)^2 + (\log y - \mu)^2 \} \right] f(\mu | D) d\mu$$

and the denominator

$$f(x, y | D) = \int \frac{1}{cx\sqrt{2\pi}} \exp\left\{ -\frac{1}{2c^2} (\log x - \mu)^2 \right\} f(\mu | D) d\mu$$

$$\times \int \frac{1}{cy\sqrt{2\pi}} \exp\left\{ -\frac{1}{2c^2} (\log y - \mu)^2 \right\} f(v | D) dv$$

where $f(\cdot | D)$ is the conditional probability density function of the population of actual bandweights, given the observed sample D. This is similar to the result in (7.3) but here the measurements are assumed to have a lognormal distribution.

The density function f could be estimated by considering the sample frequencies of the $\log z_i$'s. However, such an approach assumes D is the population rather than a sample from it and ignores sampling error. Instead *kernel density estimation* (Section 7.3) is used. In this context a *kernel function* K_i is fitted about the logarithm of data point $(\log z_i)$ for $i = 1, \ldots, n$. The density function at a particular point z_i will have mean $\log z_i$ and variance equal to the estimated variance of the data set; the function is $N(\log z_i, c^2)$. Further smoothing of the distribution can be achieved by the introduction of a smoothing parameter λ in the variance of the kernel function such that

$$f(\mu | D) = \frac{1}{n} \sum_{i=1}^{n} \frac{1}{\lambda c \sqrt{2\pi}} \exp\left\{ -\frac{1}{2\lambda^2 c^2} (\mu - \log z_i)^2 \right\}.$$

The smoothing parameter λ should be larger for smaller sample sizes to account for sampling variability. With this expression for $f(\mu | D)$, V can be written as

$$V = \frac{Q_2(m) \exp\{ -(s/c)^2 \}}{Q_1(\log x) Q_1(\log y)} \tag{8.7}$$

where

$$Q_j(u) = \frac{1}{n} \sum_{i=1}^{n} \frac{1}{\sqrt{1 + j\lambda^2}} \exp\left\{ -\frac{j}{2c^2} \frac{(u - \log z_i)^2}{1 + j\lambda^2} \right\}$$

for $j = 1, 2$ (Berry, 1991a) and u should be replaced by m, $\log x$ or $\log y$ as appropriate.

Some comments regarding the choice of the smoothing parameter λ are made by Berry (1991a). If x and y are close to each other in a region in which there are few of the z_i's then the choice of λ is crucial; V will be larger with smaller values of λ. Large values of λ smooth out the gaps in the density estimate with a corresponding decrease in the peaks (since the integral of the density estimate over the whole range has to be 1). Rare values of x and y close together will provide small values of the denominator and large values of V. If λ is large these rare values of x and y will not provide such small values of the denominator and thus not such large values of V. This will be beneficial to the defence. Conversely, common values of x and y close together will provide relatively large values of the denominator and smaller values of V. A large value of λ will tend to increase the value of V which will not be beneficial to the defence. Berry (1991a) suggests that the band weights of the crime sample (or the suspect sample, but not both since they may both come from the same person) be included in D. This will have the effect of making V smaller and thus weakening the evidence against the suspect.

All of the above applies to the case of a single crime band and a single suspect band. If the two bands produced by a single locus probe can be assumed independent, a value of V may be obtained for the two pairs of bands. Let (y_1, y_2) be the crime sample measurements and (x_1, x_2) be the suspect sample measurements. If the suspect and crime samples come from the same person then x_1 is paired with either y_1 or y_2. Let V_{jk} be the value of V for pair $\{(x_j, y_k); j, k = 1, 2.\}$ then V for the two pairs is

$$V = \tfrac{1}{2} V_{11} V_{22} + \tfrac{1}{2} V_{12} V_{21}.$$

If $x_1 \gg x_2$ and $y_1 \gg y_2$, with respect to measurement standard deviations, then V_{12} and V_{21} are small and

$$V \simeq \tfrac{1}{2} V_{11} V_{22}.$$

If both the crime and suspect samples are reasonably homozygotic then $x_1 = x_2$ and $y_1 = y_2$. All four single-band likelihood ratios $\{V_{jk}; j, k = 1, 2.\}$ are equal and

$$V = V_{11}^2.$$

For several independent single locus probes, the likelihood ratios for the individual probes are calculated as above and then multiplied together.

8.5.3   Likelihood ratio for bivariate data

An approach for assessing DNA profiles considered as a pair of band positions in which measurement errors are correlated, using an approach based on likelihood ratios, was presented by Berry *et al.* (1992). The full calculation of V requires numerical integration of a bivariate convolution of a Normal density and a kernel density estimate; it is not reproduced here.

8.6   EXAMPLE OF LIKELIHOOD RATIO (NEW YORK v. CASTRO)

This is discussed at length in Berry (1991a). José Castro, an Hispanic male, was accused of murdering Vilma Ponce in 1987. He was identified as the murderer by the victim's common law husband (Lewin, 1989). DNA evidence centred round a blood stain found on Castro's watch which was compared with the blood of Vilma Ponce. Lifecodes Corporation reported odds of 1:189 200 000. (See Lander, 1989, for criticisms of the Lifecodes report.) Lifecodes used as an estimate of the standard deviation of the measurement error 0.6% of the mean of the molecular weight (measured in kilobases) of the crime sample and the suspect sample. (Notice here the confusion which may be caused by the terminology 'crime' and 'suspect' sample. The source or bulk form is that from Vilma Ponce. The receptor or transferred particle form is that on Castro's watch. While it is reasonable to think of the crime sample as being the sample found on Castro's watch, it does not seem reasonable to think of Vilma Ponce's blood as being 'suspect'.)

The results for two probes, D2S44 and D17S79, are given in Table 8.10. Here, D17S79 gives results for two bands but D2S44 gives a result for a single band only. The single band may have occurred for one of several reasons as given in Section 8.2.1.

Table 8.10   Band lengths (*kb*) in the Castro case

Sample	Probe		
	D2S44	D17S79	D17S79
Ponce (Source)	10.162	3.869	3.464
Watch (Receptor)	10.350	3.877	3.541
Difference	0.188	0.008	0.077
Mean	10.256	3.873	3.503
S.d. (0.6% of mean)	0.0615	0.0232	0.0210
3 s.d	0.1845	0.0696	0.0630
Match decision	No	Yes	No

Table 8.11 Frequency distribution of D17S79 alleles for Hispanic database from Figure 1(C) of Balazs *et al.*, (1989). Cell widths equal 0.05 *kb*

Allele size (*kb*)	Frequency	Allele size (*kb*)	Frequency
2.65	2	3.85	19
2.70	1	3.90	7
2.85	1	3.95	8
2.90	1	4.00	15
3.05	2	4.05	6
3.15	1	4.10	6
3.20	3	4.15	5
3.25	6	4.20	3
3.30	14	4.25	2
3.35	22	4.30	2
3.40	30	4.40	2
3.45	14	4.55	1
3.50	17	4.60	1
3.55	24	4.65	1
3.60	27	4.80	1
3.65	24	4.85	1
3.70	7	5.00	1
3.75	8	5.30	1
3.80	9	Total = 295	

A match is declared if the source form lies within three standard deviations of the receptor form. Despite two of the results giving a decision that there was no match, Lifecodes claimed a match in all three cases (Berry, 1991a).

Lifecodes databases for these two probes are given in Balazs *et al.* (1989). The 'population' frequencies for these two probes (assuming that all the bands match) were estimated using bins of width six standard deviations, corresponding to a match criterion of three standard deviations, centred on the victim's band measurements where the standard deviation was 0.6% of the mean molecular weight. The 'population' frequencies were determined from databases for 295 Hispanics (for probe D17S79) and for 292 Hispanics (for probe D2S44), given in Balazs *et al.* (1989) and repeated in Tables 8.11 and 8.12. The required frequencies were estimated by using the database proportions which lie within three standard deviations or 1.8% of the *victim's* band weights and are given in Table 8.13. Thus, for example, for D2S44 the bin is 10.162 ± 0.183 ($0.018 \times 10.162 = 0.183$). If D2S44 is assumed to be a true homozygote than the match probability for these two probes is

$$0.049^2 \times (2 \times 0.111 \times 0.155) = 8.26 \times 10^{-5} \simeq 1/12\,000.$$

If D2S44 is taken to be an apparent homozygote then a conservative estimate for

Table 8.12 Frequency distribution of D2S44 alleles for Hispanic database from Figure 3(C) of Balazs *et al.*, (1989). Cell widths equal 0.10 *kb* up to 11 *kb* and 0.20 *kb* from 11 *kb*

Allele size (*kb*)	Frequency	Allele size (*kb*)	Frequency
7.8	2	10.6	9
7.9	2	10.7	17
8.0	1	10.8	35
8.1	1	10.9	26
8.2	1	11.0	15
8.3	2	11.2	19
8.4	1	11.4	12
8.5	3	11.6	10
8.6	1	11.8	12
8.7	2	12.0	16
9.5	1	12.2	16
9.6	1	12.4	14
9.7	1	12.6	12
9.8	2	12.8	11
9.9	1	13.0	6
10.0	2	13.2	1
10.1	3	13.4	4
10.2	6	13.8	1
10.3	4	14.2	1
10.4	7	14.4	1
10.5	8	15.0	2
		Total = 292	

Table 8.13 Band lengths (*kb*) and match probability calculations based on source sample in the Castro case

	Probe		
	D2S44	D17S79	D17S79
Ponce	10.162	3.869	3.464
Bin	10.162 ± 0.183	3.869 ± 0.070	3.464 ± 0.062
Estimated population frequency	0.049	0.111	0.155

the match probability is

$$(2 \times 0.049) \times (2 \times 0.111 \times 0.155) = 3.37 \times 10^{-3} \simeq 1/300.$$

The above calculations are scene-anchored (which is correct for correspondence probabilities—see comments on transfer evidence in Section 5.2—but does not

Table 8.14 Band lengths (*kb*) and match probability calculations based on receptor sample in the Castro case.

	Probe		
	D2S44	D17S79	D17S79
Watch	10.350	3.877	3.541
Bin	10.350 ± 0.186	3.877 ± 0.070	3.541 ± 0.064
Estimated population frequency	0.079	0.111	0.191

Table 8.15 Values of V for Castro case from (8.6). Ponce's band weight $= x$, weight of watch sample $= y$

Probe	Band length (*kb*)	Identifier	Standard deviation c			
			0.0042	0.006	0.010	0.020
D2S44	$x = 10.162$ $y = 10.350$	V_1	0.289	2.20	5.46	3.96
D17S79	$x = 3.869$ $y = 3.877$	V_2	20.0	14.1	9.19	4.97
D17S79	$x = 3.464$ $y = 3.541$	V_3	0.013	0.304	1.60	2.11
$V = V_o$, D2S44 homozygous			0.011	10	220	82
$V = V_e$, D2S44 heterozygous			0.019	2.4	20	10

agree with Weir and Gaut's (1993) principle of anchoring on the receptor sample). If frequencies are based on the transferred-particle form the results given in Table 8.14 are obtained. The probabilities are $(0.079^2 \times 2 \times 0.111 \times 0.191)$ equals $2.65 \times 10^{-4} \simeq 1/3800$ for a true homozygote and $((2 \times 0.079) \times (2 \times 0.111 \times 0.191))$ equals $6.70 \times 10^{-3} \simeq 1/150$ for an apparent homozygote, different from those based on the source sample.

8.6.1 Estimation of the likelihood ratio V

The likelihood ratios V calculated from (8.6) for each of the three bands from D2S44 and D17S79 for various values of the standard deviation c are given in Table 8.15. The relevant data sets are taken from Balazs *et al.* (1989). The source is taken to be Ponce with measurement x. The receptor (transferred-particle form) is taken to be Castro's watch. For D2S44 assumed homozygous, $V_o = \frac{1}{2} V_2 V_3 V_1^2$; for D2S44 assumed heterozygous, $V_e = \frac{1}{2} V_2 V_3 (\frac{1}{2} V_1)$. For example, for $c = 0.0042$,

Table 8.16   Values of V for Castro case from (8.7). Ponce's band weight $= x$, weight of watch sample $= y$

Probe	Band length (*kb*)	Standard deviation c and smoothing parameter λ				
		$c = 0.0042$ $\lambda = 1$	0.006 1	0.006 5	0.010 1	0.010 2
D2S44	$x = 10.162$ $y = 10.350$	0.305	2.56	1.97	6.78	5.86
D17S79	$x = 3.869$ $y = 3.877$	17.7	13.9	16.1	9.34	9.74
D17S79	$x = 3.464$ $y = 3.541$	0.017	0.343	0.343	1.62	1.68
V_o: D2S44 homozygous		0.014	15	10	350	280
V_e: D2S44 heterozygous		0.022	3.1	2.7	26	24

$V_o = \frac{1}{2} \times 20 \times 0.013 \times 0.289^2 = 0.011$, $V_e = \frac{1}{2} \times 20 \times 0.013 \times \frac{1}{2} \times 0.289 = 0.019$. Table 8.16 shows the likelihood ratio calculated from (8.7) for each of the three bands from D2S44 and D17S79 for various values of the standard deviation c and smoothing parameter λ. The relevant data sets are again taken from Balazs *et al.* (1989). The source is taken to be Ponce with measurement x. The receptor (transferred-particle form) is taken to be Castro's watch.

8.6.2   Comments

The relevant data sets used are Hispanic databases from Balazs *et al.* (1989), Figures 1C and 3C, reproduced in Berry (1991a). The suspect, Castro, was a Hispanic. As has been discussed earlier, it is wrong to assume the race of the suspect as the reference population. The appropriate population is that which may have provided the blood on the watch. However, Lifecodes used the Hispanic database, Berry (1991a) did and it is used here also.

The interpretation of V when it is less than one is of interest. Consider Table 8.16, $c = 0.0042$, $\lambda = 1$ and D2S44 homozygous: $V_o = 0.014$, $V_o^{-1} \simeq 70$. Then it can be said that the evidence of the blood is about 70 times more likely if the blood came from the relevant population than if it came from Ponce.

The difference in probability distributions for DNA profile band weights for different ethnic groups may be used to provide information about the ethnic group of the donor of an unknown crime stain (Evett *et al.*, 1992a). Consider a comparison between hypotheses:

- C: the sample came from a Caucasian,
- $\bar{C}$: the sample came from an Afro-Caribbean.

Table 8.17 Probability p that a relative has the same genotype as the suspect and the corresponding value for V assuming allelic frequencies of 0.1, from Weir and Hill (1993) (Reproduced by permission of The Forensic Science Society)

Suspect	Relative	Probability p	Likelihood ratio, V
A_iA_j	Father or son	$(p_i + p_j)/2$	10
	Full brother	$(1 + p_i + p_j + 2p_ip_j)/4$	6.67
	Half brother	$(p_i + p_j + 4p_ip_j)/4$	16.67
	Uncle or nephew	$(p_i + p_j + 4p_ip_j)/4$	16.67
	First cousin	$(p_i + p_j + 12p_ip_j)/8$	25
	Unrelated	$2p_ip_j$	50
A_iA_i	Father or son	p_i	10
	Full brother	$(1 + p_i)^2/4$	3.3
	Half brother	$p_i(1 + p_i)/2$	18.2
	Uncle or nephew	$p_i(1 + p_i)/2$	18.2
	First cousin	$p_i(1 + 3p_i)/4$	30.8
	Unrelated	p_i^2	100

Let $E = (x_1, x_2)$ denote the band weights from the DNA profile from the crime stain. The likelihood ratio is then

$$\frac{Pr(E|C)}{Pr(E|\overline{C})} = \frac{f(x_1, x_2|C)}{f(x_1, x_2|\overline{C})}$$

where the change from Pr to f is made to emphasise that the band weights are continuous and that the ratio is one of probability density functions. Notice, also, that this is an example where the conditioning events are not complementary. The probability density function for (x_1, x_2) is taken to be the product of the probability density functions for each of x_1 and x_2 separately, assuming Hardy–Weinberg equilibrium. These separate probability density functions may be estimated by kernel density estimation using information from suitable databases of Caucasian and Afro-Caribbean band weights. A value of the likelihood ratio greater than one will provide support for a Caucasian origin, a value of the likelihood ratio less than one will provide support for an Afro-Caribbean origin. If more than one probe is used and data are available the corresponding likelihood ratios can be multiplied together, assuming linkage equilibrium. Further details are available in Evett *et al.* (1992a).

8.7   RELATED INDIVIDUALS

The discussion so far has assumed that the reference population contains no related individuals. Unrelated individuals have a very low probability of sharing the same profile but the probability increases for related individuals. Consider the two hypotheses under consideration to be:

- C: the suspect left the crime sample;
- $\bar{C}$: a relative of the suspect left the crime sample.

Let the evidence be the measurements (x_1, x_2) of the source sample (suspect) and (y_1, y_2) of the receptor sample (crime). Assume that, if C is true, that the numerator of the likelihood ratio is 1, an assumption which will not necessarily be true but will serve to illustrate the point. Let (y_1, y_2) correspond to alleles A_i, A_j with population frequencies p_i and p_j. If there were no familial relationship between the source and receptor samples then the denominator would be $2p_ip_j$ for $i \neq j$ and p_i^2 for $i = j$. The effects of different familial relationships are given in Table 8.17 from Weir and Hill (1993) where numerical values are given assuming allelic frequencies of 0.1. The numerator is assumed to have the value 1. See also Brookfield (1994) for further examples.

8.7.1   More than two hypotheses

Consider a situation in which the evidence E is that of a DNA profile of a stain of body fluid found at the scene of a crime and of the DNA profile from a suspect which matches in some sense the crime stain and that there are three hypotheses to be considered:

- C: the suspect left the crime stain;
- $\bar{C}_1$: a random member of the population left the crime stain;
- $\bar{C}_2$: a brother of the defendant left the crime stain.

This situation has been discussed by Evett (1992). Let p_0, p_1 and p_2 denote the prior probabilities for each of these three hypotheses ($p_0 + p_1 + p_2 = 1$). Assume that $Pr(E|C) = 1$. Denote $Pr(E|\bar{C}_1)$ by q_1 and $Pr(E|\bar{C}_2)$ by q_2. Also, $\bar{C}$, the complement of C, is the conjunction of $\bar{C}_1$ and $\bar{C}_2$. Then, using Bayes' Theorem,

$$Pr(C|E) = \frac{Pr(E|C)p_0}{Pr(E|C)p_0 + Pr(E|\bar{C}_1)p_1 + Pr(E|\bar{C}_2)p_2}$$

$$= \frac{p_0}{p_0 + q_1p_1 + q_2p_2},$$

$$Pr(\bar{C}|E) = \frac{q_1p_1 + q_2p_2}{p_0 + q_1p_1 + q_2p_2},$$

and hence the posterior odds in favour of C are

$$\frac{p_0}{q_1p_1 + q_2p_2}.$$

Let the relevant population size be N and let the number of siblings be n where $N \gg n$. It can be assumed that $p_0 = 1/N$, $p_1 = (N-n)/N$ and $p_2 = (n-1)/N$. The posterior odds in favour of C are approximately equal to

$$\frac{1}{q_1 N + q_2(n-1)}.$$

Example 8.1   Evett (1992) gives an example where two single-locus probes are used, each giving two bands which match the suspect. Each of the four bands has a frequency of 0.01 so the probability (q_1) of a match from a random individual is $(4 \times 0.01^4) = 1/25\,000\,000$. If $N = 100\,000$, and there are no siblings so that $n = 0$ then the posterior odds in favour of C are $1/(Nq_1) = 250$. If there is a brother, then $q_2 \simeq (1/4)^2$, $n = 2$ and the posterior odds are

$$1 \left/ \left\{ \frac{1}{250} + \left(\frac{1}{4}\right)^2 \right\} \right. = \frac{1}{0.004 + 0.063} \simeq 15.$$

The existence of the brother has reduced the posterior odds by a factor of over 15 from 250 to 15.

Notice that the consideration of three hypotheses has led to consideration of posterior odds in favour of one of the hypotheses. The determination of a likelihood ratio under such circumstances is discussed in Section 5.1.2. The likelihood ratio is used for comparing hypotheses in pairs. For comparing more than two hypotheses, hypotheses have to be combined in some meaningful way to provide two hypotheses for comparison.

8.8   SUMMARY

There are compelling reasons why the likelihood ratio is preferable to a match-binning approach. First, there is no dichotomy between a match and otherwise. All comparisons will provide a value for the likelihood ratio, some will be greater than 1, lending support to a common origin, others will be less than 1, lending support to a different origin. Second, the likelihood ratio is easy to interpret and is easily incorporated with other evidence. The frequency result produced by match-binning is open to the confusion attached to coincidence probability approaches to evidence assessment described in Section 4.6.

The NRC Report (1992, p. 85) in a short comment dismissed the likelihood ratio approach (Section 8.4.1). However, Berry (1991a) and Berry *et al.* (1992) provide methods, based on sound statistical techniques, for determining the distribution of fragment positions. It is difficult to see why the results of the likelihood ratio techniques could not be understood easily by a layman. The match-binning approach can lead to considerable confusion. In contrast to this, a likelihood ratio approach has a very intuitive interpretation as has been mentioned often in this book. Of course, the statistical methodology is complicated but a statistical layman does not need to understand the methodology in the same way as he is not expected to understand the methodology used in the construction of a DNA profile.

In conclusion, various methods for assessing the value of DNA profiling have been discussed: match-binning with fixed and floating bins and bin width equal to the match criterion, and bin width bigger than the match criterion, guessing, the ceiling principle and likelihood ratio approaches. Of these, the likelihood ratio approach is the one which most succeeds in providing an evaluation of evidence which is continuously related to the relative band positions and which is easy to interpret.

8.9   APPENDIX

8.9.1   Constancy of variance of logarithm of band weights

Assume the mean and variance of band weight x are μ and $c^2\mu^2$, so that the standard deviation is proportional to the mean band weight. The Taylor expansion of $\log x$ about μ gives

$$\log x \simeq \log \mu + (x - \mu)\frac{1}{\mu} + \cdots$$

The mean of $\log x$ is approximately equal to $\log \mu$. The variance of $\log x$ is approximately equal to the mean of $(\log x - \log \mu)^2$ and, hence, the mean of $(x - \mu)^2/\mu^2$. However, the mean of $(x - \mu)^2$ is just the variance of x, which is $c^2\mu^2$. Thus the variance of $\log x$ is c^2, a constant independent of the band weight.

8.9.2   Derivation of V for population data

From Berry (1991a), the DNA evidence is the measurements of a single band weight for a crime scene sample (y) and a suspect sample (x). There are also available band weights ($z_1, \ldots, z_n$) from some relevant population. Also

$$\log y \sim N(v, c^2),$$
$$\log x \sim N(\mu, c^2),$$
$$\log z_i \sim N(\mu_i, c^2), \qquad i = 1, \ldots, n.$$

Then

$$f(x, y, z_1, \ldots, z_n | \mu = v) = \frac{1}{2\pi c^2 xy} \int \exp\left\{\frac{1}{2c^2}[(\log x - \mu)^2 + (\log y - \mu)^2]\right\} d\mu$$

$$\times \prod_{i=1}^{n} \frac{1}{c z_i \sqrt{2\pi}} \exp\left\{-\frac{1}{2c^2}(\log z_i - \mu_i)^2\right\} d\mu_i.$$

Now,

$$\int \frac{1}{cz_i\sqrt{2\pi}} \exp\left\{ -\frac{1}{2c^2}(\log z_i - \mu_i)^2 \right\} d\mu_i = \frac{1}{z_i} \int N(\log z_i, c^2) d\mu_i = \frac{1}{z_i},$$

where the notation $\int N(\log z_i, c^2) d\mu_i$ denotes the integral of μ_i. Normally distributed with mean $\log z_i$ and variance c^2, over the range of μ_i; this integral is equal to 1 (see Chapter 3).

Note also that $\frac{1}{2}\{(\log x - \mu)^2 + (\log y - \mu)^2\}$ may be written as $s^2 + (m - \mu)^2$ where $s^2 = (\log x - \log y)^2/4$ and $m = (\log x + \log y)/2$. Thus,

$$f(x, y, z_1, \ldots, z_n | \mu = v) = \frac{1}{2\pi c^2 xy} \int \exp\left\{ -\frac{1}{c^2}[s^2 + (m - \mu)^2] \right\} d\mu \times \prod_{i=1}^{n} \frac{1}{z_i}$$

Now,

$$\frac{1}{2\pi c^2 xy} \int \exp\left\{ -\frac{1}{c^2}[s^2 + (m - \mu)^2] \right\} d\mu$$

$$= \frac{1}{2cxy\sqrt{\pi}} \exp\left(-\frac{s^2}{c^2} \right) \int \frac{1}{c\sqrt{\pi}} \exp\left\{ -\frac{1}{c^2}(\mu - m)^2 \right\} d\mu$$

$$= \frac{1}{2cxy\sqrt{\pi}} \exp\left(-\frac{s^2}{c^2} \right) \int N\left(m, \frac{c^2}{2} \right) d\mu$$

$$= \frac{1}{2cxy\sqrt{\pi}} \exp\left(-\frac{s^2}{c^2} \right)$$

since the term in the integral is 1. Thus

$$f(x, y, z_1, \ldots, z_n | \mu = v) = \frac{1}{2cxy\sqrt{\pi}} \exp\left(-\frac{s^2}{c^2} \right) \prod_{i=1}^{n} \frac{1}{z_i}$$

$$= K \times \exp\left(-\frac{s^2}{c^2} \right)$$

where

$$K = \frac{1}{2cxy\sqrt{\pi}} \prod_{i=1}^{n} \frac{1}{z_i}.$$

The denominator of V is

$$f(x, y, z_1, \ldots, z_n | \bar{C})$$

$$= \sum_{i=1}^{n} f(x, y, z_1, \ldots, z_n | \text{person } i \text{ is guilty}) Pr(\text{person } i \text{ is guilty} | \bar{C})$$

$$= \frac{1}{n} \sum_{i=1}^{n} f(x, y, z_1, \ldots, z_n | \mu_i = v)$$

$$= \frac{K}{n} \sum_{i=1}^{n} \exp\left(-\frac{s^2}{c^2} \right)$$

where it is assumed that $Pr(\text{person } i \text{ is guilty} | \bar{C}) = 1/n$; i.e., that all n people in the population are assigned an equal probability, $1/n$, of guilt. The final line follows by analogy with the argument for the numerator. The ratio of the numerator to the denominator then gives the result that

$$V = \frac{\exp\{-(s/c)^2\}}{n^{-1} \sum_{i=1}^{n} \exp\{-(s_i/c)^2\}}.$$

References

Aitchison, J. and Dunsmore, I. R. (1975) *Statistical Prediction Analysis*, Cambridge University Press, Cambridge.

Aitken, C. G. G. (1986) Statistical discriminant analysis in forensic science. *Journal of the Forensic Science Society*, **26**, 237–247.

Aitken, C. G. G. (1991a) Report on International Conference on Forensic Statistics. *Journal of the Royal Statistical Society, Series A*, **154**, 45–48. Selected papers included in pp. 49–130.

Aitken, C. G. G. (1991b) Populations and samples. In *The Use of Statistics in Forensic Science* (Eds. C. G. G. Aitken and D. A. Stoney), pp. 51–82, Ellis Horwood, Chichester.

Aitken, C. G. G. (1993) Statistics and the law: report of a discussion session at The Royal Statistical Society Conference, Sheffield, September, 1992. *Journal of the Royal Statistical Society, Series A*, **156**, 301–304.

Aitken, C. G. G. and MacDonald, D. G. (1979) An application of discrete kernel methods to forensic odontology. *Applied Statistics*, **28**, 55–61.

Aitken, C. G. G. and Gammerman, A. (1989) Probabilistic reasoning in evidential assessment. *Journal of the Forensic Science Society*, **29**, 303–316.

Aitken, C. G. G. and Robertson, J. (1987) A contribution to the discussion of probabilities and human hair comparisons. *Journal of Forensic Sciences*, **32**, 684–689.

Aitken, C. G. G. and Stoney, D. A. (Editors) (1991) *The Use of Statistics in Forensic Science*, Ellis Horwood, Chichester.

Anderson, T. W. (1984) *An Introduction to Multivariate Analysis*, Wiley, New York.

Baird, M., Balazs, I., Giusti, A., Miyazaki, L., Nicholas, L., Wexler, K., Kanter, E., Glassberg, J., Allen, F., Rubenstein, P. and Sussman, L. (1986) Allele frequency distribution of two highly polymorphic DNA sequences in three ethnic groups and its application to the determination of paternity. *American Journal of Human Genetics*, **39**, 489–501.

Balazs, I., Baird, M., Clyne, M. and Meade, E. (1989) Human population genetic studies of five hypervariable DNA loci. *American Journal of Human Genetics*, **44**, 182–190.

Balding, D. J. and Donnelly, P. (1994) Inference in forensic identification (with discussion). *Journal of the Royal Statistical Society, Series A*, to appear.

Barnard, G. A. (1958) Thomas Bayes—a biographical note (together with a reprinting of Bayes, 1764). *Biometrika*, **45**, 293–315. Reprinted in Pearson and Kendall (1970), 131–153.

Bates, J. W. and Lambert, J. A. (1991) Use of the hypergeometric distribution for sampling in forensic glass comparison. *Journal of the Forensic Science Society*, **31**, 449–455.

Bayes, T. (1764) An essay towards solving a problem in the doctrine of chances. *Philosophical Transactions of the Royal Society of London for 1763*, **53**, 370–418. Reprinted with Barnard (1958) in Pearson and Kendall (1970), 131–153.

Bentham, J. (1827) *Rationale of Judicial Evidence, Specially Applied to English Practice*. (Ed. J. S. Mill) Hunt and Clarke, London.

Berger, J. O. and Sellke, T. (1987) Testing a point null hypothesis: the irreconcilability of P values and evidence. *Journal of the American Statistical Association*, **82**, 112–139.

Bernoulli, J. (1713) *Ars Conjectandi*, Basle, Switzerland.
Bernoulli, N. (1709) *Specimina Artis Conjectandi ad quaestiones juris applicatae*, Basle, Switzerland.
Berry, D. A. (1991a) Inferences using DNA profiling in forensic identification and paternity cases. *Statistical Science*, **6**, 175–205.
Berry, D. A. (1991b) Probability of paternity. In *The Use of Statistics in Forensic Science* (Eds. C. G. G. Aitken and D.A. Stoney), pp. 150–156, Ellis Horwood, Chichester.
Berry, D. A. and Geisser. S. (1986) Inference in cases of disputed paternity. In *Statistics and the Law* (Eds. M. H. DeGroot, S. E. Fienberg and J. B. Kadane) Wiley, New York, pp. 353–382.
Berry, D. A., Evett, I. W. and Pinchin, R. (1992) Statistical inference in crime investigations using deoxyribonucleic acid profiling (with discussion). *Applied Statistics*, **41**, 499–531.
Briggs, T. J. (1978) The probative value of bloodstains on clothing. *Medicine, Science and the Law*, **18**, 79–83.
Brookfield, J. F. Y. (1994) The effect of relatives on the likelihood ratio associated with DNA profile evidence in criminal cases. *Journal of the Forensic Science Society*, **34**, 193–197.
Brown, G. A. and Cropp, P. L. (1987) Standardised nomenclature in forensic science. *Journal of the Forensic Science Society*, **27**, 393–399.
Buckleton, J. S., Walsh, K. A. J., Seber, G. A. F. and Woodfield, D. G. (1987) A stratified approach to the compilation of blood group frequency surveys. *Journal of the Forensic Science Society*, **27**, 103–112.
Buckleton, J. S. and Walsh, K. A. J. (1991) Knowledge-based systems. In *The Use of Statistics in Forensic Science* (Eds. C. G. G. Aitken and D. A. Stoney), pp. 186–206, Ellis Horwood, Chichester.
Buckleton, J. S., Walsh, K. A. J. and Evett, I. W. (1991) Who is 'random man'? *Journal of the Forensic Science Society*, **31**, 463–468.
Budowle, B., Giusti, A. M., Waye, J. S., Baechtel, F. S., Fourney, R. M., Adams, D. E., Presley, L. A., Deadman, H. A. and Monson, K. L. (1991) Fixed-bin analysis for statistical evaluation of continuous distributions of allele data from VNTR loci, for use in forensic comparisons. *American Journal of Human Genetics*, **48**, 841–855.
Chakraborty, R. and Kidd, K. K. (1991) The utility of DNA typing in forensic work. *Science*, **254**, 1735–1739.
Chan, K. P. S. and Aitken, C. G. G. (1989) Estimation of the Bayes' factor in a forensic science problem. *Journal of Statistical Computation and Simulation*, **33**, 249–264.
Cohen, J. E. (1990) DNA fingerprints for forensic identification: potential effects on data interpretation of subpopulation heterogeneity and band number volatility. *American Journal of Human Genetics*, **46**, 358–368.
Cohen, L. J. (1977) *The Probable and the Provable*, Clarendon Press, Oxford.
Cohen, L. J. (1988) The difficulty about conjunction in forensic proof. *The Statistician*, **37**, 415–416.
Coleman, R. F. and Walls, H. J. (1974) The evaluation of scientific evidence. *Criminal Law Review*, 276–287.
Condorcet, Le Marquis de (1785) *Essai sur l'application de l'analyse à la probabilité des décisions rendues à la pluralité des voix*, Imprimerie Royale, Paris.
Cournot, A. A. (1838) Sur les applications du calcul des chances à la statistique judiciaire. *Journal des mathematiques pures et appliques*, **3**, 257–334.
Cullison, A. D. (1969) Probability analysis of judicial fact–finding: a preliminary outline of the subjective approach. *University of Toledo Law Review*, **1969**, 538–598.
Curnow, R. N. and Wheeler, S. (1993) Probabilities of incorrect decisions in paternity cases using multilocus deoxyribonucleic acid probes. *Journal of the Royal Statistical Society, Series A*, **156**, 207–223.

Dabbs, M. G. D. and Pearson, E. F. (1970) Heterogeneity in glass. *Journal of the Forensic Science Society*, **10**, 139–148.

Dabbs, M. G. D. and Pearson, E. F. (1972) Some physical properties of a large number of window glass specimens. *Journal of Forensic Sciences*, **17**, 70–78.

Darroch, J. (1987) Probability and criminal trials; some comments prompted by the Splatt trial and The Royal Commission. *The Professional Statistician*, **6**, 3–7.

Davis, R. J. and DeHaan, J. D. (1977) A survey of men's footwear. *Journal of the Forensic Science Society*, **17**, 271–285.

Dawid, A. P. (1987) The difficulty about conjunction. *The Statistician*, **36**, 91–97.

Dawid, A. P. (1994) The island problem: coherent use of identification evidence. In *Aspects of Uncertainty: A Tribute to D. V. Lindley* (Eds. P. R. Freeman and A. F. M. Smith), Wiley, Chichester, pp. 159–170.

Dawson, T. C. (1993) DNA profiling: evidence for the prosecution. *Journal of the Forensic Science Society*, **33**, 238–242.

De Groot, M. H. (1970) *Optimal statistical decisions*, McGraw–Hill Book Company, New York, U.S.A.

Devlin, B. and Risch, N. J. (1992) Ethnic differentiation at VNTR loci with special reference to forensic applications. *American Journal of Human Genetics*, **51**, 534–548.

Devlin, B., Risch, N. J. and Roeder, K. (1992) Forensic inference from DNA fingerprints. *Journal of the American Statistical Association*, **87**, 337–350.

Devlin, B., Risch, N. J. and Roeder, K. (1993) Statistical evaluation of DNA fingerprinting: a critique of the NRC report. *Science*, **259**, 748–749, 837.

Diaconis, P. and Freedman, D. (1981) The persistence of cognitive illusions. *The Behavioural and Brain Sciences*, **4**, 333–334.

Edwards, W., Lindman, H. and Savage, L. J. (1963) Bayesian statistical inference for psychological research. *Psychological Review*, **70**, 193–242. Reprinted in *Robustness of Bayesian analyses* (Ed. J. Kadane), Elsevier, Amsterdam, Netherlands, 1984.

Eggleston, R. (1983) *Evidence, Proof and Probability (2nd edition)*. Weidenfeld and Nicolson, London.

Essen–Möller, E. (1938) Die Biesweiskraft der Ähnlichkeit im Vater Schaftsnachweis; Theoretische Grundlagen. *Mitt. Anthrop. Ges. (Wein)*, **68**, 598.

Evett, I. W. (1977) The interpretation of refractive index measurements. *Forensic Science*, **9**, 209–217.

Evett, I. W. (1978) The interpretation of refractive index measurements, II. *Forensic Science International*, **12**, 34–47.

Evett, I. W. (1983) What is the probability that this blood came from that person? A meaningful question? *Journal of the Forensic Science Society*, **23**, 35–39.

Evett, I. W. (1984) A quantitative theory for interpreting transfer evidence in criminal cases. *Applied Statistics*, **33**, 25–32.

Evett, I. W. (1986) A Bayesian approach to the problem of interpreting glass evidence in forensic science casework. *Journal of the Forensic Science Society*, **26**, 3–18.

Evett, I. W. (1987a) Bayesian inference and forensic science: problems and perspectives. *The Statistician*, **36**, 99–105.

Evett, I. W. (1987b) On meaningful questions: a two–trace transfer problem. *Journal of the Forensic Science Society*, **27**, 375–381.

Evett, I. W. (1992) Evaluating DNA profiles in the case where the defence is 'It was my brother'. *Journal of the Forensic Science Society*, **32**, 5–14.

Evett, I. W. (1993a) Establishing the evidential value of a small quantity of material found at a crime scene. *Journal of the Forensic Science Society*, **33**, 83–86.

Evett, I. W. (1993b) Criminalistics: the future of expertise. *Journal of the Forensic Science Society*, **33**, 173–178.

Evett, I. W. and Lambert, J. A. (1982) The interpretation of refractive index measurements, III. *Forensic Science International,* **20,** 237–245.

Evett, I. W. and Lambert, J. A. (1984) The interpretation of refractive index measurements, IV. *Forensic Science International,* **26,** 149–163.

Evett, I. W. and Lambert, J. A. (1985) The interpretation of refractive index measurements, V. *Forensic Science International,* **27,** 97–110.

Evett, I. W., Cage, P. E. and Aitken, C. G. G. (1987) Evaluation of the likelihood ratio for fibre transfer evidence in criminal cases. *Applied Statistics,* **36,** 174–180.

Evett, I. W., Werrett, D. J., Gill, P. and Buckleton, J. S. (1989a) DNA fingerprinting on trial. *Nature,* **340,** 435.

Evett, I. W., Werrett, D. J. and Smith, A. F. M. (1989b) Probabilistic analysis of DNA profiles. *Journal of the Forensic Science Society,* **29,** 191–196.

Evett, I. W., Werrett, D. J. and Buckleton, J. S. (1989c) Paternity calculations from DNA multilocus profiles. *Journal of the Forensic Science Society,* **29,** 249–254.

Evett, I. W. and Buckleton, J. (1990) The interpretation of glass evidence. A practical approach. *Journal of the Forensic Science Society,* **30,** 215–223.

Evett, I. W., Buffery, C., Willott, G. and Stoney, D. (1991) A guide to interpreting single locus profiles of DNA mixtures in forensic cases. *Journal of the Forensic Science Society,* **31,** 41–47.

Evett, I. W. and Weir, B. S. (1992) Flawed reasoning in court. *Chance,* **4,** 19–21.

Evett, I. W., Pinchin, R. and Buffery, C. (1992a) An investigation of the feasibility of inferring ethnic origin from DNA profiles. *Journal of the Forensic Science Society,* **32,** 301–306.

Evett, I. W., Scranage, J. and Pinchin, R. (1992b) An efficient statistical procedure for interpreting DNA single locus profiling data in crime cases. *Journal of the Forensic Science Society,* **32,** 307–326.

Evett, I. W., Scranage, J. and Pinchin, R. (1993) An illustration of the advantages of efficient statistical methods for RFLP analysis in forensic science. *American Journal of Human Genetics,* **52,** 498–505.

Fairley, W. B. (1973) Probabilistic analysis of identification evidence. *Journal of Legal Studies,* **II,** 493–513.

Fairley, W. B. and Mosteller, W. (1977) *Statistics and Public Policy.* Addison–Wesley, London, pp. 355–379.

Fienberg, S. E. (Editor) (1989) *The Evolving Role of Statistical Assessments as Evidence in the Courts,* Springer-Verlag, New York.

Fienberg, S. E. and Kadane, J. B. (1983) The presentation of Bayesian statistical analyses in legal proceedings. *The Statistician,* **32,** 88–98.

Fienberg, S. E. and Schervish, M. J. (1986) The relevance of Bayesian inference for the presentation of statistical evidence and for legal decision making. *Boston University Law Review,* **66,** 771–798.

Fienberg, S. E. and Finkelstein, M. P. (1995) Bayesian statistics and the law. In *Bayesian Statistics 5* (Eds. J. M. Bernardo, J. O. Berger, A. P. Dawid and A. F. M. Smith), Oxford University Press (to appear).

Findlay, D. R. (1993) DNA profiling and criminal law—a merger or a takeover. *Journal of the Forensic Science Society,* **33,** 234–237.

Finkelstein, M. O. and Fairley, W. B. (1970) A Bayesian approach to identification evidence. *Harvard Law Review,* **83,** 489–517.

Finkelstein, M. O. and Fairley, W. B. (1971) A comment on 'Trial by mathematics'. *Harvard Law Review,* **84,** 1801–1809.

Finkelstein, M. O. and Levin, B. (1990) *Statistics for Lawyers,* Springer Verlag, New York.

Finney, D. J. (1977) Probabilities based on circumstantial evidence. *Journal of the American Statistical Association,* **72,** 316–318.

Fisher, R. A. (1951) Standard calculations for evaluating a blood group system. *Heredity*, **5**, 51–102.

Fong, W. and Inami, S. H. (1986) Results of a study to determine the probability of chance match occurrences between fibres known to be from different sources, *Journal of Forensic Sciences*, **31**, 65–72.

Freeling, A. N. S. and Sahlin, N.-E. (1983) Combining evidence. In *Evidentiary Value*, (Eds. P. Gardenfors, B. Hansson and N.-E. Sahlin), C. W. K. Gleerup, Lund, Sweden, 58–74.

Gaensslen, R. E., Bell, S. C. and Lee, H. C. (1987a) Distribution of genetic markers in United States populations: 1. Blood group and secretor systems. *Journal of Forensic Sciences*, **32**, 1016–1058.

Gaensslen, R. E., Bell, S. C. and Lee, H. C. (1987b) Distribution of genetic markers in United States populations: 2. Isoenzyme systems. *Journal of Forensic Sciences*, **32**, 1348–1381.

Gaensslen, R. E., Bell, S. C. and Lee, H. C. (1987c) Distribution of genetic markers in United States populations: 3. Serum group systems and haemoglobin variants. *Journal of Forensic Sciences*, **32**, 1754–1774.

Garber, D. and Zabell, S. (1979) On the emergence of probability. *Archive for History of Exact Sciences*, **21**, 33–53.

Gastwirth, J. L. (1988a) *Statistical Reasoning in Law and Public Policy, Volume 1: Statistical concepts and Issues of Fairness*, Statistical modelling and decision science series (Eds. G. L. Lieberman and I. Olkin), Academic Press, San Diego.

Gastwirth, J. L. (1988b) *Statistical Reasoning in Law and Public Policy, Volume 2: Tort Law, Evidence and Health*, Statistical modelling and decision science series (Eds. G. L. Lieberman and I. Olkin), Academic Press, San Diego.

Gaudette, B. D. (1982) A supplementary discussion of probabilities and human hair comparisons. *Journal of Forensic Sciences*, **27**, 279–289.

Gaudette, B. D. and Keeping, E. S. (1974) An attempt at determining probabilities in human scalp hair comparison. *Journal of Forensic Sciences*, **19**, 599–606.

Gelfand, A. E. and Solomon, H. (1973) A study of Poisson's models for jury verdicts in criminal and civil trials. *Journal of the American Statistical Association*, **68**, 271–278.

Gettinby, G. (1984) An empirical approach to estimating the probability of innocently acquiring bloodstains of different *ABO* groups in clothing. *Journal of the Forensic Science Society*, **24**, 221–227.

Gettinby, G., Peterson, M. and Watson, N. (1993) Statistical interpretation of DNA evidence. *Journal of the Forensic Science Society*, **33**, 212–217.

Gjertson, D. W., Mickey, M. R., Hopfield, J., Takenouchi, T. and Terasaki, P. (1988) Calculations of the probability of paternity using DNA sequences. *American Journal of Human Genetics*, **43**, 860–869.

Good, I. J. (1950) *Probability and the Weighing of Evidence*, Charles Griffin and Company Limited, London.

Good, I. J. (1956) 'Discussion of paper by G. Spencer Brown' in *Information Theory: Third London Symposium, 1955*, (Ed. C. Cherry), pp. 13–14, Butterworths, London.

Good, I. J. (1991) Weight of evidence and the Bayesian likelihood ratio In *The Use of Statistics in Forensic Science* (Eds. C. G. G. Aitken and D. A. Stoney), Ellis Horwood, Chichester, pp. 85–106.

Grieve, M. C. and Dunlop, J. (1992) A practical aspect of the Bayesian interpretation of fibre evidence. *Journal of the Forensic Science Society*, **32**, 169–175.

Groom, P. S. and Lawton, M. E. (1987) Are they a pair? *Journal of the Forensic Science Society*, **27**, 189–192.

Grove, D. M. (1980) The interpretation of forensic evidence using a likelihood ratio. *Biometrika*, **67**, 243–246.

Grove, D. M. (1981) The statistical interpretation of refractive index measurements. *Forensic Science International*, **18**, 189–194.

Grove, D. M. (1984) The statistical interpretation of refractive index measurements *II* : the multiple source problem. *Forensic Science International*, **24**, 173–182.

Grubb, A. (1993) Legal aspects of DNA profiling. *Journal of the Forensic Science Society*, **33**, 228–233.

Grunbaum, B. W., Selvin, S., Pace, N. and Block, D. M. (1978) Frequency distribution and discrimination probability of twelve genetic variants in human blood as functions of race, sex and age. *Journal of Forensic Sciences*, **23**, 577–587.

Grunbaum, B. W., Selvin, S., Myrhe, B. A. and Pace, N. (1980) Distribution of gene frequencies and discrimination probabilities for 22 human blood genetic systems in four racial groups. *Journal of Forensic Sciences*, **25**, 428–444.

Habbema, J. D. F., Hermans, J. and van den Broek, K. (1974) A stepwise discrimination program using density estimation. In *Compstat 1974* (Ed. G. Bruckman), Physica Verlag, Vienna, pp. 100–110.

Hacking, I. (1975) *The Emergence of Probability*, Cambridge University Press, Cambridge.

Harrison, P. H., Lambert, J. A. and Zoro, J. A. (1985) A survey of glass fragments recovered from clothing of persons suspected of involvement in crime. *Forensic Science International*, **27**, 171–187.

Harvey, W., Butler, O., Furness, J. and Laird, R. (1968) The Biggar murder: dental, medical, police and legal aspects. *Journal of the Forensic Science Society*, **8**, 155–219.

Hrechak, A. K. and McHugh, J. A. (1990) Automated fingerprint recognition using structural matching. *Pattern Recognition*, **23**, 893–904.

Hummel, K. (1971) Biostatistical opinion of parentage based upon the results of blood group tests. in *Biostatistische Abstammungsbegutachtung mit Blutgruppenbefunden* (Ed. P. Schmidt), Gustav Fisher, Stuttgart, Germany. (Quoted in *Family Law Quarterly*, 1976, **10**, 262.)

Jarjoura, D., Jamison, J. and Androulakakis, S. (1994) Likelihood ratios for deoxyribonucleic (DNA) typing in criminal cases. *Journal of Forensic Sciences*, **39**, 64–73.

Jeffreys, A. J. (1993) DNA typing: approaches and applications. *Journal of the Forensic Science Society*, **33**, 204–211.

Jeffreys, A. J., Wilson, V. and Thein, S. L. (1985a) Hypervariable 'minisatellite' regions in human DNA. *Nature*, **314**, 67–73.

Jeffreys, A. J., Wilson, V. and Thein, S. L. (1985b) Individual specific 'fingerprints' of human DNA. *Nature*, **316**, 76–79.

Jeffreys, A. J., Wilson, V. and Morton, D. B. (1987) DNA fingerprints of dogs and cats. *Animal Genetics*, **18**, 1–15.

Jeffreys, H. (1983) *Theory of Probability*, 3rd edition. Clarendon Press, Oxford.

Jones, D. A. (1972) Blood samples: probability of discrimination. *Journal of the Forensic Science Society*, **12**, 355–359.

Kaye, D. H. (1993a) *Proceedings of the Second International Conference on Forensic Statistics*. Arizona State University, Center for the Study of Law, Science and Technology, Tempe, Arizona, U.S.A. Selected papers included in *Jurimetrics*, **34(1)**, pp. 1–115.

Kaye, D. H. (1993b) DNA evidence: probability, population genetics and the courts. *Harvard Journal of Law and Technology*, **7**, 101–172.

Kaye, D. H. and Aickin, M. (1986) *Statistical methods in discrimination litigation*, Marcel Dekker Inc., New York.

Kendall, M. G. and Buckland, W. R. (1982) *A Dictionary of Statistical Terms (4th edition)*, Longman, London.

Kind, S. S. (1994) Crime investigation and the criminal trial: a three chapter paradigm of evidence. *Journal of the Forensic Science Society*, **34**, 155–164.

Kind, S. S., Wigmore, R., Whitehead, P.H. and Loxley, D.S. (1979) Terminology in forensic science. *Journal of the Forensic Science Society*, **19**, 189–192.

Kingston, C. R. (1964) Probabilistic analysis of partial fingerprint patterns. D. Crim. dissertation, University of California, Berkeley, U.S.A.

Kingston, C. R. (1965a) Applications of probability theory in criminalistics. *Journal of the American Statistical Association*, **60**, 70–80.

Kingston, C. R. (1965b) Applications of probability theory in criminalistics—II. *Journal of the American Statistical Association*, **60**, 1028–1034.

Kingston, C. R. (1988) Discussion of 'A critical analysis of quantitative fingerprint individuality models'. *Journal of Forensic Sciences*, **33**, 9–11.

Kirk, P. L. and Kingston, C. R. (1964) Evidence evaluation and problems in general criminalistics. Presented at the Sixteenth Annual Meeting of the American Academy of Forensic Sciences. Chicago, Illinois, in a symposium *The Principles of Evidence Evaluation*.

Koehler, J. J. (1993) Error and exaggeration in the presentation of DNA evidence at trial. *Jurimetrics*, **34**, 21–39.

Kuo, M. (1982) Linking a bloodstain to a missing person by genetic inheritance. *Journal of Forensic Sciences*, **27**, 438–444.

Laing, D. K., Hartshorne, A. W. and Harwood, R. J. (1986) Colour measurements on single textile fibres. *Forensic Science International*, **30**, 65–77.

Laing, D. K., Hartshorne, A. W., Cook, R. and Robinson, G. (1987) A fibre data collection for forensic scientists—collection and examination methods. *Journal of Forensic Sciences*, **32**, 364–369.

Lambert, J. A. and Evett, I. W. (1984) The refractive index distribution of control glass samples examined by the forensic science laboratories in the United Kingdom. *Forensic Science International*, **26**, 1–23.

Lander, E. S. (1989) DNA fingerprinting on trial. *Nature*, **339**, 501–505.

Lander, E. S. (1991) Research on DNA typing catching up with courtroom application. *American Journal of Human Genetics*, **48**, 819–823.

LaPlace, Le Marquis de (1886) Essai philosophique sur les probabilitiés. *Introduction to Theorie Analytique des Probabilités*, Oeuvres Complètes de LaPlace, Gauthier–Villars, Paris.

Lempert, R. (1977) Modelling relevance. *Michigan Law Review*, **89**, 1021–1057.

Lempert, R. (1991) Some caveats concerning DNA as criminal identification evidence: with thanks to the Reverend Bayes. *Cardozo Law Review*, **13**, 303–341.

Lenth, R. V. (1986) On identification by probability. *Journal of the Forensic Science Society*, **26**, 197–213.

Lewin, R. (1989) DNA typing on the witness stand. *Science*, **244**, 1033–1035.

Lewontin, R. C. (1993) Which population? (Letter) *American Journal of Human Genetics*, **52**, 205.

Lewontin, R. C. and Hartl, D. C. (1991) Population genetics in forensic DNA typing. *Science*, **254**, 1745–1750.

Lindley, D. V. (1957) A statistical paradox. *Biometrika*, **44**, 187–192. (Comments by M. S. Bartlett and M. G. Kendall appear in volume 45, 533–534.)

Lindley, D. V. (1977a) A problem in forensic science. *Biometrika*, **64**, 207–213.

Lindley, D. V. (1977b) Probability and the law. *The Statistician*, **26**, 203–212.

Lindley, D. V. (1980) L. J. Savage—his work in probability and statistics. *The Annals of Statistics*, **8**, 1–24.

Lindley, D. V. (1987) The probability approach to the treatment of uncertainty in artificial intelligence and expert systems. *Statistical Science*, **2**, 17–24.

Lindley, D. V. (1991) Probability. In *The Use of Statistics in Forensic Science* (Eds. C. G. G. Aitken and D. A. Stoney), pp. 27–50, Ellis Horwood, Chichester.

Lindley, D. V. and Scott, W. F. (1984) *New Cambridge Elementary Statistical Tables*, Cambridge University Press, Cambridge.

Mardia, K. V., Kent, J. T. and Bibby, J. M. (1979) *Multivariate Analysis*, Academic Press, London.

Mardia, K. V., Li, Q. and Hainsworth,T. J. (1992) On the Penrose hypothesis on fingerprint patterns. *IMA Journal of Mathematics Applied in Medicine and Biology*, **9**, 289–294.

McQuillan, J. and Edgar, K. (1992) A survey of the distribution of glass on clothing. *Journal of the Forensic Science Society*, **32**, 333–348.

Meier, P. and Zabell, S. (1980) Benjamin Peirce and the Howland will. *Journal of the American Statistical Association*, **75**, 497–506.

Mode, E. B. (1963) Probability and criminalistics. *Journal of the American Statistical Association*, **58**, 628–640.

Morris, J. W. and Brenner, C. H. (1988) Bloodstain classification errors revisited. *Journal of the Forensic Science Society*, **28**, 49–53.

Morris, J. W., Sanda, A. I. and Glassberg, J. (1989) Biostatistical evaluation of evidence from continuous allele frequency distribution deoxyribonucleic acid (DNA) probes in reference to disputed paternity and identity. *Journal of Forensic Sciences*, **34**, 1311–1317.

Mourant, A. E., Kope, A. C. and Domaniewska-Sobczak, K. (1958) *The ABO Blood Groups*, Blackwell, Oxford.

National Research Council (1992) *DNA Technology in Forensic Science*, National Academy Press, Washington, D.C., U.S.A.

Ogino, C. and Gregonis, D. J. (1981) Indirect typing of a victim's blood using paternity testing. Presentation before the California Association of Criminalists 57-th Semi-annual Seminar, Pasadena, California.

Osterburg, J. W., Parasarathy, T., Raghaven, T. E. S. and Sclove, S. L. (1977) Development of a mathematical formula for the calculation of fingerprint probabilities based on individual characteristics. *Journal of the American Statistical Association*, **72**, 772–778.

Owen, G. W. and Smalldon, K. W. (1975) Blood and semen stains on outer clothing and shoes not related to crime: report of a survey using presumptive tests. *Journal of Forensic Sciences*, **20**, 391–403.

Parker, J. B. (1966) A statistical treatment of identification problems. *Forensic Science Society Journal*, **6**, 33–39.

Parker, J. B. (1967) The mathematical evaluation of numerical evidence. *Forensic Science Society Journal*, **7**, 134–144.

Parker, J. B. and Holford, A. (1968) Optimum test statistics with particular reference to a forensic science problem. *Applied Statistics*, **17**, 237–251.

Peabody, A. J., Oxborough, R. J., Cage, P. E. and Evett, I. W. (1983) The discrimination of cat and dog hairs. *Journal of the Forensic Science Society*, **23**, 121–129.

Pearson, E. F., May, R. W. and Dabbs, M. G. D. (1971) Glass and paint fragments found in men's outer clothing—report of a survey. *Journal of Forensic Sciences*, **16**, 283–300.

Pearson, E. S. and Kendall, M. G. (Editors) (1970) *Studies in the History of Statistics and Probability*. Charles Griffin, London.

Poisson, S. D. (1837) Recherches sur la probabilité des jugements en matière criminelle et en matière civile, précédées des règles générales du calcul des probabilités. Bachelier, Imprimeur-Libraire, Paris.

Pounds, C. A. and Smalldon, K. W. (1978) The distribution of glass fragments in front of a broken window and the transfer of fragments to individuals standing nearby. *Journal of the Forensic Science Society*, **18**, 197–203.

Rabinovitch, N. L. (1969) Studies in the history of probability and statistics XXII: probability in the Talmud. *Biometrika*, **56**, 437–441.

Rabinovitch, N. L. (1973) *Probability and Statistical Inference in Ancient and Medieval Jewish Literature*, University of Toronto Press, Toronto.

Risch, N. J. and Devlin, B. (1992) On the probability of matching DNA fingerprints. *Science*, **255**, 717–720.

Roeder, K. (1994) DNA fingerprinting: a review of the controversy (with discussion). *Statistical Science*, **9**, 222–278.

Rothwell, T. J. (1993) DNA profiling and crime investigation—the European context. *Journal of the Forensic Science Society*, **33**, 226–227.

Salmon, D. and Salmon, C. (1980) Blood groups and genetic markers polymorphisms and probability of paternity. *Transfusion*, **20**, 684–694.

Schatkin, S. B. (1984) *Disputed Paternity Proceedings, Volumes I and II, 4th Edition, Revised*, Matthew Bender, New York.

Sclove, S. L. (1979) The occurrence of fingerprint characteristics as a two-dimensional process. *Journal of the American Statistical Association*, **74**, 588–595.

Sclove, S. L. (1980) The occurrence of fingerprint characteristics as a two-dimensional Poisson process. *Commun. Statist.—Theor. Meth.*, **A9(7)**, 675–695.

Seheult, A. (1978) On a problem in forensic science. *Biometrika*, **65**, 646–648.

Selvin, S., Grunbaum, B. W. and Myhre, B. A. (1983) The probability of non-discrimination or likelihood of guilt of an accused: criminal identification. *Journal of the Forensic Science Society*, **23**, 27–33.

Selvin, S. and Grunbaum, B. W. (1987) Genetic marker determination in evidence bloodstains: the effect of classification errors on probability of non-discrimination and probability of concordance. *Journal of the Forensic Science Society*, **27**, 57–63.

Sensabaugh, G. (1984) Summary blood typing, defendant's exhibit *E*, Evidentiary hearing, *P. v. Young*, Circuit Court for the County of Alpena, State of Michigan, May 23rd, 1984.

Shafer, G. (1976) *A Mathematical Theory of Evidence*, Princeton University Press, Princeton.

Shafer, G. (1978) Non-additive probabilities in the work of Bernoulli and Lambert. *Archive for History of Exact Sciences*, **19**, 309–370.

Shafer, G. (1982) Lindley's paradox (with discussion). *Journal of the American Statistical Association*, **77**, 325–351.

Shannon, C. E. (1948) A mathematical theory of communication. *Bell System Technical Journal*, **27**, 379–423.

Sheynin, O. B. (1974) On the prehistory of the theory of probability. *Archive for History of Exact Sciences*, **12**, 97–141.

Silverman, B. W. (1986) *Density estimation*, Chapman and Hall, London.

Simon, R. J. and Mahan, L. (1971) Quantifying burdens of proof. *Law and Society Review*, **5**, 319–330.

Simpson, E. H. (1949) Measures of diversity. *Nature*, **163**, 688.

Smalldon, K. W. and Moffat, A. C. (1973) The calculation of discriminating power for a series of correlated attributes. *Journal of the Forensic Science Society*, **13**, 291–295.

Smith, R. L. and Charrow, R. P. (1975) Upper and lower bounds for the probability of guilt based on circumstantial evidence. *Journal of the American Statistical Association*, **70**, 555–560.

South Carolina Law Enforcement Division, DNA Analysis Laboratory (1992) Procedures for the detection of restriction fragment length polymorphisms in human DNA, Columbia, South Carolina.

Stedman, R. (1972) Human population frequencies in twelve blood grouping systems. *Journal of the Forensic Science Society*, **12**, 379–413.

Stedman, R. (1985) Blood group frequencies of immigrant and indigenous populations from South East England. *Journal of the Forensic Science Society*, **25**, 95–134.

Stoney, D. A. (1984) Statistics applicable to the inference of a victim's blood type from familial testing. *Journal of the Forensic Science Society*, **24**, 9–22.

Stoney, D. A. (1991) Transfer evidence *in The Use of Statistics in Forensic Science* (Eds. C. G. G. Aitken and D. A. Stoney), Ellis Horwood, Chichester, pp. 107–138.

Stoney, D. A. (1994) Relaxation of the assumption of relevance and an application to one-trace and two-trace problems. *Journal of the Forensic Science Society*, **34**, 17–21.

Stoney, D. A. and Thornton, J. I. (1986) A critical analysis of quantitative fingerprint individuality models. *Journal of Forensic Sciences*, **33**, 11–13.

Thompson, W. C. and Schumann, E. L. (1987) Interpretation of statistical evidence in criminal trials. The prosecutor's fallacy and the defence attorney's fallacy. *Law and Human Behaviour*, **11**, 167–187.

Thompson, Y. and Williams, R. (1991) Blood group frequencies of the population of Trinidad and Tobago, West Indies. *Journal of the Forensic Science Society*, **31**, 441–447.

Tippett, C. F., Emerson, V. J., Fereday, M. J., Lawton, F. and Lampert, S. M. (1968) The evidential value of the comparison of paint flakes from sources other than vehicles. *Journal of the Forensic Science Society*, **8**, 61–65.

Tribe, L. (1971) Trial by mathematics: precision and ritual in the legal process. *Harvard Law Review*, **84**, 1329–1393.

Tryhorn, F. G. (1935) The assessment of circumstantial scientific evidence. *Police Journal*, **8**, 401–411.

Tversky, A. and Kahneman, D. (1974) Judgement under uncertainty: heuristics and biases. *Science*, **185**, 1124–1131.

Vito, G. F. and Latessa, E. J. (1989) *Statistical Applications in Criminal Justice*, Law and Criminal Justice Series, volume 10, Sage Publications, London.

Wakefield, J. C., Skene, A. M., Smith, A. F. M. and Evett, I. W. (1991) The evaluation of fibre transfer evidence in forensic science: a case study in statistical modelling. *Applied Statistics*, **40**, 461–476.

Walsh, K. A. J. and Buckleton, J. S. (1986) On the problem of assessing the evidential value of glass fragments embedded in footwear. *Journal of the Forensic Science Society*, **26**, 55–60.

Walsh, K. A. J. and Buckleton, J. S. (1988) A discussion of the law of mutual independence and its application to blood group frequency. *Journal of the Forensic Science Society*, **28**, 95–98.

Walsh, K. A. J. and Buckleton, J. S. (1991) Calculating the frequency of occurrence of a blood type for a 'random man'. *Journal of the Forensic Science Society*, **31**, 49–58.

Walsh, K. A. J. and Buckleton, J. S. (1994) Assessing prior probabilities considering geography. *Journal of the Forensic Science Society*, **34**, 47–51.

Weir, B. S. (1992) Independence of VNTR alleles defined as floating bins. *American Journal of Human Genetics*, **51**, 992–997.

Weir, B. S. and Evett, I. W. (1992) Whose DNA? *American Journal of Human Genetics*, **50**, 869.

Weir, B. S. and Evett, I. W. (1993) Reply to Lewontin (Letter) *American Journal of Human Genetics*, **52**, 206.

Weir, B. S. and Gaut, B. S. (1993) Matching and binning DNA fragments in forensic science. *Jurimetrics*, **34**, 9–19.

Weir, B. S. and Hill, W. G. (1993) Population genetics of DNA profiles. *Journal of the Forensic Science Society*, **33**, 218–225.

Yellin, J. (1979) Book review of '*Evidence, Proof and Probability, 1st Edition*', Eggleston, R. (1978) Weidenfeld and Nicolson, London; *Journal of Economic Literature*, Pittsburgh p. 583.

Zabell, S. (1976) Book review of '*Probability and Statistical Inference in Ancient and Medieval Jewish Literature*', Rabinovitch, N. L. (1973), University of Toronto Press, Toronto, Canada; *Journal of the American Statistical Association*, **71**, 996–998.

Author Index

The contributions of authors who would otherwise be subsumed anonymously in the '*et al*' of multi-author papers has been recognised by including page references for all authors of such papers, even though only the name of the first author appears on the page itself.

Subject Index